Springer Theses

Recognizing Outstanding Ph.D. Research

Aims and Scope

The series "Springer Theses" brings together a selection of the very best Ph.D. theses from around the world and across the physical sciences. Nominated and endorsed by two recognized specialists, each published volume has been selected for its scientific excellence and the high impact of its contents for the pertinent field of research. For greater accessibility to non-specialists, the published versions include an extended introduction, as well as a foreword by the student's supervisor explaining the special relevance of the work for the field. As a whole, the series will provide a valuable resource both for newcomers to the research fields described, and for other scientists seeking detailed background information on special questions. Finally, it provides an accredited documentation of the valuable contributions made by today's younger generation of scientists.

Theses may be nominated for publication in this series by heads of department at internationally leading universities or institutes and should fulfill all of the following criteria

- They must be written in good English.
- The topic should fall within the confines of Chemistry, Physics, Earth Sciences, Engineering and related interdisciplinary fields such as Materials, Nanoscience, Chemical Engineering, Complex Systems and Biophysics.
- The work reported in the thesis must represent a significant scientific advance.
- If the thesis includes previously published material, permission to reproduce this must be gained from the respective copyright holder (a maximum 30% of the thesis should be a verbatim reproduction from the author's previous publications).
- They must have been examined and passed during the 12 months prior to nomination.
- Each thesis should include a foreword by the supervisor outlining the significance of its content.
- The theses should have a clearly defined structure including an introduction accessible to new PhD students and scientists not expert in the relevant field.

Indexed by zbMATH.

More information about this series at https://link.springer.com/bookseries/8790

Yûto Murashita

Fluctuation Theorems under Divergent Entropy Production and their Applications for Fundamental Problems in Statistical Physics

Doctoral Thesis accepted by
University of Tokyo, Tokyo, Japan

Springer

Author
Dr. Yûto Murashita
Department of Physics
The University of Tokyo
Tokyo, Japan

Supervisor
Prof. Masahito Ueda
Department of Physics
The University of Tokyo
Tokyo, Japan

ISSN 2190-5053 ISSN 2190-5061 (electronic)
Springer Theses
ISBN 978-981-16-8637-5 ISBN 978-981-16-8638-2 (eBook)
https://doi.org/10.1007/978-981-16-8638-2

This Springer imprint is published by the registered company Springer Nature Singapore Pte Ltd.
The registered company address is: 152 Beach Road, #21-01/04 Gateway East, Singapore 189721, Singapore

Supervisor's Foreword

Understanding macroscopic thermodynamic phenomena from microscopic dynamics is a grand challenge that has hovered over physicists ever since the era of Gibbs and Boltzmann. Over the past quarter century, the fluctuation theorem has played a pivotal role in unveiling several fundamental aspects of emergent irreversibility in systems far from equilibrium. In his thesis, Yuto Murashita studies the fluctuation theorem in extreme yet practically important situations with divergent entropy production and discusses its implications to a few fundamental issues in statistical mechanics.

In the standard framework of the fluctuation theorem, the entropy production is expressed in terms of the ratio of the probability of the original dynamics to that of the time-reversed one. Measure theory tells us that this ratio is not always well defined, especially when entropy production diverges and such cases belong to a new class of irreversibility, which is referred to as absolute irreversibility. Yuto extends the fluctuation theorem so as to be applicable to situations with absolute irreversibility. As a different situation with divergent entropy production, Yuto considers a system which is coupled simultaneously to multiple heat baths at different temperatures as in the case of the Feynman ratchet. In the overdamped limit, the entropy production is divergent since velocities relax not to an equilibrium state but to a nonequilibrium steady state. Despite this singular behavior of the fast degrees of freedom, the fluctuation theorem can be shown for the dynamics of slow degrees of freedom. Furthermore, Yuto applies the fluctuation theorem with absolute irreversibility to two fundamental problems in statistical physics: the Gibbs paradox and the Loschmidt paradox. The Gibbs paradox originates from the problem of gas mixing and is closely connected with the issue in statistical-mechanical problem concerning how to define entropy. Yuto shows that the fluctuation theorem with absolute irreversibility can be utilized as a definition of entropy by considering the gas-mixing process. Finally, Yuto revisits the Loschmidt paradox from the viewpoint of the fluctuation theorem with absolute irreversibility. The Loschmidt paradox concerns how irreversible behaviors emerge from reversible equations of motion. In a closed Hamiltonian system, an imperfect Loschmidt demon is invoked to define an entropy production consistent with the fluctuation theorem. Remarkably, the entropy thus defined exhibits system-specific behavior reminiscent of the Kolmogorov-Sinai entropy and features emergent irreversibility.

I believe that this thesis makes a seminal contribution toward resolving two long-standing problems in statistical physics, namely the Gibbs paradox and the Loschmidt paradox.

Tokyo, Japan
December 2021

Masahito Ueda

Parts of this thesis have been published in the following journal articles:

Y. Murashita, K. Funo and M. Ueda, "Nonequilibrium equalities in absolutely irreversible processes" Physical Review E 90, 042110 (2014). DOI: https://doi.org/10.1103/PhysRevE.90.042110. Parts of Chap. 3 of the thesis are reprinted with permission from this paper. Copyright 2014 by the American Physical Society.

Y. Murashita and M. Esposito, "Overdamped stochastic thermodynamics with multiple reservoirs" Physical Review E 94, 062148 (2016). DOI: https://doi.org/10.1103/PhysRevE.94.062148.

Y. Murashita and M. Ueda, "Gibbs Paradox Revisited from the Fluctuation Theorem with Absolute Irreversibility" Physical Review Letters 118, 060601 (2017). DOI: https://doi.org/10.1103/PhysRevLett.118.060601. Parts of Chap. 5 of the thesis are reprinted with permission from this paper. Copyright 2017 by the American Physical Society.

Y. Murashita, N. Kura and M. Ueda, "Entropy production of a closed Hamiltonian system via the detailed fluctuation relation" Physical Review Research 3, 033224 (2021). DOI: https://doi.org/10.1103/PhysRevResearch.3.033224. Parts of Chap. 6 of the thesis are reprinted from this paper under the Creative Commons Attribution 4.0 International license.

Acknowledgements

The studies in this thesis were conducted when the author was a doctoral student at Prof. Masahito Ueda's group in the University of Tokyo. I would like to express my deep sense of gratitude to my collaborators and colleagues.

First of all, I would like to express my best gratitude to my supervisor, Prof. Masahito Ueda. His inspiring comments and considerable encouragement during our frequent discussions greatly helped me to advance my research. In particular, his critical and constructive comments were indispensable to complete the studies in Chap. 5 and Chap. 6. Furthermore, he read the manuscript of the thesis with great care and gave me enormous suggestions to improve it.

I would like to thank my collaborator, Prof. Massimiliano Esposito. He accepted me as a visiting student at the University of Luxembourg through the oversea dispatch program of my graduate school. Without his extensive expertise or kind support in Luxembourg, the study in Chap. 4 would not be completed.

I am grateful for support from my subadvisor, Prof. Yasunobu Nakamura. He gave me regular opportunities to discuss experimental aspects of our work and taught me on the latest relevant experiments.

I sincerely appreciate considerable assistance from my collaborators. Dr. Naoto Kura regularly discussed with me the study in Chap. 6 and made significant contributions to its mathematical aspects. Dr. Yuta Masuyama frequently visited us to have discussions and taught me fundamentals of experiments of superconducting qubits. Dr. Ken Funo taught me quantum aspects of the fluctuation theorems and gave me enormous opportunities to discuss with. Dr. Tomohiro Shitara helped me to establish mathematical and information-theoretic aspects of our studies. Dr. Yuto Ashida gave me constructive comments from his wide viewpoint to improve impacts and accessibility of our studies. Dr. Zongping Gong taught me key references relevant to our studies on the basis of his comprehensive knowledge of the field.

I would like to express my gratitude to the former and present members of Prof. Ueda's group, especially to Prof. Shunsuke Furukawa and Mrs. Asako Takeuchi, who gave me generous support for my life in the university.

I am also indebted to the members in the University of Luxembourg, especially to Dr. Riccardo Rao, Dr. Artur Wachtel and Dr. Marko Mravlak. They helped me to live a fruitful life there.

I am grateful to Prof. Hosho Katsura and the members of his group for providing me with the office space.

Finally, I acknowledge support from the Japan Society for the Promotion of Science through a fellowship (Grant No. JP15J00410) and the Program for Leading Graduate School (MERIT).

Contents

Chapter 2
Review on Fluctuation Theorems

We here give a brief history of fluctuation theorems. The fluctuation theorems are equalities about the probability distribution functions for the thermodynamic quantities that are essentially related to the entropy production. Remarkably, they are applicable to a wide range of nonequilibrium situations. In this chapter, we first review their discovery in a shear-driven flow and their derivations to see that they are valid under various types of dynamics. Then, we consider various types of the fluctuation theorems including the Jarzynski equality and the Crooks fluctuation theorems, and discuss their general structure. Finally, we briefly discuss the fluctuation theorems under nonuniform temperatures.

2.1 Discovery of Fluctuation Theorems

We here review the discovery of the fluctuation theorems. In a reversible dissipative chaotic system, an asymptotic symmetry of the probability distribution function of the entropy production was found, which is now called the steady-state fluctuation theorem. Later in the same system in the transient setup that starts from equilibrium, the same symmetry is found in a finite-time interval, which is called the transient fluctuation theorem. These fluctuation theorems give quantitative expressions for the probability of the second-law violation. Furthermore, the fluctuation theorems can be regarded as a generalization of the linear response theory. The fluctuation theorems in this section are restricted to systems with time-independent driving.

2.1.1 Asymptotic Steady-State Fluctuation Theorem

In 1993, Evans, Cohen and Morriss [1] proposed a novel symmetry for the probability distribution function of the entropy production. Motivated by a previous research on one-dimensional systems [2], they heuristically conjectured that the measure invari-

Y. Murashita, *Fluctuation Theorems under Divergent Entropy Production and their Applications for Fundamental Problems in Statistical Physics*, Springer Theses, https://doi.org/10.1007/978-981-16-8638-2_2

ant under dynamics that describes a nonequilibrium steady state is asymptotically written as

$$\mu_\tau[\Gamma_i] = \frac{\exp[-\tau \sum_{\lambda_n[\Gamma_i]>0} \lambda_n[\Gamma_i]]}{\sum_j \exp[-\tau \sum_{\lambda_n[\Gamma_j]>0} \lambda_n[\Gamma_j]]}, \tag{2.1}$$

where Γ_i represents the ith trajectory $\Gamma_{i,t}$ in phase space from time $t = 0$ to τ, and $\lambda_n[\Gamma_i]$'s are the local Lyapunov exponents. Note that the sum is now taken over all positive Lyapunov exponents. Suppose that we can construct an "anti-segment" $\bar{\Gamma}_i$ so that it may be in some sense a reversal of the original segment Γ_i and $\bar{\Gamma}_i$ may satisfy the time reversal of the original equations of motion. Then, the sign of the Lyapunov exponents are reversed as

$$\lambda_n[\bar{\Gamma}_i] = -\lambda_n[\Gamma_i]. \tag{2.2}$$

Consequently, the measure for the reversed trajectory can be written as

$$\begin{aligned} \mu_\tau[\bar{\Gamma}_i] &= \frac{\exp[-\tau \sum_{\lambda_n[\bar{\Gamma}_i]>0} \lambda_n[\bar{\Gamma}_i]]}{\sum_j \exp[-\tau \sum_{\lambda_n[\Gamma_j]>0} \lambda_n[\Gamma_j]]} \\ &= \frac{\exp[\tau \sum_{\lambda_n[\Gamma_i]<0} \lambda_n[\Gamma_i]]}{\sum_j \exp[-\tau \sum_{\lambda_n[\Gamma_j]>0} \lambda_n[\Gamma_j]]}. \end{aligned} \tag{2.3}$$

Therefore, the probability ratio of Γ_i to $\bar{\Gamma}_i$ is given by

$$\frac{\mu_\tau[\Gamma_i]}{\mu_\tau[\bar{\Gamma}_i]} = \exp[Nd\langle\alpha[\Gamma_i]\rangle_\tau \tau], \tag{2.4}$$

where we define the time-averaged phase-space contraction rate per a degree of freedom by

$$\langle\alpha[\Gamma_i]\rangle_\tau = -\frac{1}{Nd}\sum_n \lambda_n[\Gamma_i] \tag{2.5}$$

with N and d being the number of particles and the dimension of space, respectively. Note that the sum is now taken over all Lyapunov exponents, and is not restricted to positive exponents. Since the phase-space contraction rate changes its sign under the reversal operation, the probability distribution of the contraction rate $P(\alpha)$ asymptotically has the following symmetry

$$\begin{aligned} P(\alpha_\tau) &= \sum_{\Gamma_i:\langle\alpha[\Gamma_i]\rangle_\tau=\alpha_\tau} \mu_\tau[\Gamma_i] \\ &= \exp[Nd\alpha_\tau\tau] \sum_{\Gamma_i:\langle\alpha[\Gamma_i]\rangle_\tau=\alpha_\tau} \mu_\tau[\bar{\Gamma}_i] \\ &= \exp[Nd\alpha_\tau\tau]P(-\alpha_\tau), \end{aligned} \tag{2.6}$$

or equivalently,

$$\frac{P(\alpha_\tau)}{P(-\alpha_\tau)} = \exp[Nd\alpha_\tau \tau], \tag{2.7}$$

which is now called the steady-state fluctuation theorem.

To verify this symmetry of the probability distribution function, the authors of Ref. [1] conducted nonequilibrium molecular dynamic simulations of a planar shear-driven flow, namely, the Couette flow. Suppose that the fluid is sandwiched between two planes perpendicular to the y axis and one of the plane moves at a constant speed along the x axis. Then, the flow has the gradient of the x component of the local velocity u in the y direction, which we denote as $\gamma = \partial u_x/\partial y$. The flow is modeled by the following deterministic and reversible equations of motion with an isoenergetic thermostat called the SLLOD equations [3]:

$$\dot{q}_j = \frac{p_j}{m} + \mathbf{i}\gamma y_j, \tag{2.8}$$

$$\dot{p}_j = F_j - \mathbf{i}\gamma p_{j,y} - \alpha p_j, \tag{2.9}$$

where $j = 1, \ldots, N$ is the label to specify the particles and $p_j = m\dot{q}_j - \mathbf{i}mu_x(q_j)$ is the relative momentum of the jth particle with respect to the local fluid velocity $u_x(q_j) = \gamma y_j$; $\mathbf{i}$ is the unit vector in the x direction. These equations are called SLLOD, since their generator is the transpose of so-called Doll's equations [3]. The value of α is determined so that the energy of the system

$$H_0 = \sum_j \frac{p_j^2}{2m} + \Phi(q_1, \ldots, q_N) \tag{2.10}$$

may be fixed as

$$0 = \frac{dH_0}{dt} = -\alpha \sum_j \frac{p_j^2}{m} - \gamma P_{xy} V, \tag{2.11}$$

where we define the pressure tensor by

$$P_{xy} = \frac{1}{V} \sum_{j=1}^{N} \frac{p_{jx} p_{jy}}{m} - \frac{1}{2V} \sum_{j \neq j'} (x_{j'} - x_j) F_{jj',y}, \tag{2.12}$$

with $F_{jj'}$ being the force on j by j'. Therefore, α is given by

$$\alpha = -\frac{1}{Nd} \beta \gamma P_{xy} V, \tag{2.13}$$

where β is the inverse temperature defined by the kinetic energy: $1/\beta := (1/Nd) \sum_j p_j^2/m$. We can regard $\gamma P_{xy} V$ as the work done on the system by the shear driving per one degree of freedom. Since the system is isoenergetically thermostatted, we can consider that the same amount of heat is transferred to the heat reservoir. Therefore, we can physically interpret α to be the entropy production rate per one degree of freedom.

Meanwhile, the phase-space contraction rate is calculated by

$$\sum_j \frac{\partial \dot{q}_j}{\partial q_j} + \sum_j \frac{\partial \dot{p}_j}{\partial p_j} = -Nd\alpha. \tag{2.14}$$

Therefore, α in the SLLOD equations is nothing but the instantaneous phase-space compression factor per one degree of freedom. Consequently, the time average of α asymptotically satisfies the relation (2.7).

They conducted numerical simulations and obtained the probability distribution function of the time-averaged pressure tensor in a steady state during time τ. From this probability distribution function, they calculated the logarithm of the probability ratio defined by

$$\Pi(P_{xy\tau}) := \frac{1}{Nd\tau} \ln \frac{P(P_{xy\tau})}{P(-P_{xy\tau})}. \tag{2.15}$$

As a result, the value of Π was shown to coincide with α_τ, which indicates that the steady-state fluctuation theorem (2.7) holds.

An important implication of the fluctuation theorem is that the second law of thermodynamics is probabilistically violated. The violation probability is exponentially small compared with the probability of the corresponding second-law satisfying event as indicated by Eq. (2.7). When the number of the degrees of freedom Nd is small and the time interval τ is short, the second-law violation occurs relatively frequently. Meanwhile, in the long time or in large systems, the second-law violation rarely occurs due to the exponential suppression.

Another important point is that the fluctuation theorem is valid even far from equilibrium. Before the fluctuation theorem, little general equalities had been known for genuine nonequilibrium systems beyond the linear-response regime. However, the fluctuation theorem applies to a wide range of systems far from equilibrium as well as systems near equilibrium.

Two years later than the discovery, the steady-state fluctuation theorem was proven by Gallavotti and Cohen [4, 5] in a dissipative reversible chaotic system by assuming the chaotic hypothesis. The chaotic hypothesis reads [5]

> A reversible many-particle system in a stationary state can be regarded as a transitive Anosov system for the purpose of computing the macroscopic system.

An Anosov system is a paradigmatic chaotic system with well-behaved stable and unstable manifolds. Technically speaking, a dynamical system is called an Anosov system iff it satisfies transversality, covariance and hyperbolicity (see, e.g., Ref. [6]). In this system, when the initial state is randomly chosen in the Liouville measure (i.e., the uniform measure) of phase space, it is guaranteed that there exists the probability measure along the unstable manifold that describes asymptotic statistics of the stationary state, which is called the Sinai-Ruelle-Bowen (SRB) measure [7–9].[1] In fact, in a transitive Anosov system, we can coarse-grain the unstable manifold into Markov partitions, which are the tilings naturally constructed from the dynamical map of the system. Then, we can attribute each partition with a weight inversely proportional to the absolute value of the determinant of the Jacobian of the dynamical map restricted on the unstable manifold. The measure thus-constructed can be shown to describe the long-time statistics of the system. In this way, the steady-state measure essentially equivalent to Eq. (2.1) can be verified and the fluctuation theorem can therefore be shown in systems satisfying the chaotic hypothesis. We note that the chaotic hypothesis can be regarded as the nonequilibrium analog of the ergodic hypothesis, which describes the long-time statistics of the system relaxing to equilibrium.

2.1.2 *Finite-Time Transient Fluctuation Theorem*

In 1994, another type of the fluctuation theorem was found by Evans and Searles [10]. They investigated the same reversible dissipative thermostatted equations of motion (2.8) and (2.9) in a different situation. Specifically, they prepared the system in a microcanonical equilibrium without external driving, and they then switch on the driving at time $t = 0$ and drive the system during a time interval τ. In this transient setup, the following equality holds in a finite time interval (not restricted to the long-time limit):

$$\frac{P(\alpha_\tau)}{P(-\alpha_\tau)} = \exp[Nd\alpha_\tau\tau], \tag{2.16}$$

which is called the transient fluctuation theorem.

The proof of this theorem can essentially be summarized as follows [10, 11]. Let $\delta V_0(\alpha_\tau)$ denote an infinitesimal phase-space region with the phase-space contraction rate between α_τ and $\alpha_\tau + d\alpha_\tau$. After the time evolution in an interval τ, the phase-space volume contracts to be

$$\delta V_\tau(\alpha_\tau) = \exp[-Nd\alpha_\tau\tau]\delta V_0(\alpha_\tau). \tag{2.17}$$

Suppose that a trajectory Γ has the phase-space contraction rate α_τ. Then, we can construct a reversed trajectory $\bar{\Gamma}$ from an initial phase-space point obtained by a reversal operation of the endpoint of Γ. Since the reversed trajectory $\bar{\Gamma}$ has the contraction rate $-\alpha_\tau$, we obtain $\delta V_0(-\alpha_\tau) = \delta V_\tau(\alpha_\tau)$. Consequently, we obtain the ratio of the phase-space volumes as

[1] The existence of the SRB measure is mathematically shown for a transitive Anosov system [7], an axiom A diffeomorphism [8], and an axiom A flow [9]. Therefore, the chaotic hypothesis can be weakened by the replacement of a transitive Anosov system by an axiom A system.

$$\frac{\delta V_0(\alpha_\tau)}{\delta V_0(-\alpha_\tau)} = \exp[Nd\alpha_\tau \tau]. \tag{2.18}$$

Since we assume that the initial state is a microcanonical state, the probability is proportional to the phase-space volumes. Hence, the transient fluctuation theorem (2.16) holds. Note that this proof applies to a finite value of τ unlike that of the steady-state fluctuation theorem, which is valid only in the long-time limit. We also note that, even when noncontiguous regions in phase space have the same value of the phase-space compression rate, a similar discussion leads to the fluctuation theorem [12].

The transient fluctuation theorem was verified by numerical simulations of the SLLOD equations in a transient setup [10].

2.1.3 Later Studies

By later studies, it was revealed that the fluctuation theorem can be regarded as an extension of previously known response theories. In Ref. [13], it was shown that the fluctuation theorem gives a restriction for the cumulant generating function of the entropy production. As a result, in the zero-field limit, the Green-Kubo formula [14–17] and Onsager's reciprocity [18, 19] were derived. Thus, the fluctuation theorem encompasses the general relations in the linear response regime. Moreover, it was shown that the nonlinear response theory known as the Kawasaki formalism [20] can conveniently be derived from the fluctuation theorem [21].

As we have discussed, the fluctuation theorems were originally derived in an isoenergetically thermostatted system. In Ref. [22], the derivation was generalized to various types of thermostatted systems including isokinetic dynamics. They defined the dissipative function by

$$\Omega_\tau \tau = \int_0^\tau dt\ \Omega(\Gamma_t) = \ln \frac{f_0(\Gamma_0)}{f_0(\Gamma_\tau)} - \int_0^\tau dt\ \Lambda(\Gamma_t), \tag{2.19}$$

where f_0 is the probability distribution function over phase space at the initial time and Λ is the phase-space compression rate. The transient and steady-state fluctuation theorems for the dissipative function were derived. When we specify the dynamics, the dissipative function Ω_τ reduces to a quantity that essentially corresponds to the entropy production. At this stage, it was generally believed that the thermostatting mechanism is essential for the validity of the fluctuation theorem. However, in Ref. [23], the fluctuation theorem was shown to be valid even for Hamiltonian systems.

The fluctuation theorems were proven in stochastic systems as well as deterministic systems. In Ref. [24], the fluctuation theorems were derived in the Langevin dynamics. They considered a symmetry of the time-evolution operator for the probability distribution function and demonstrated that this symmetry leads to the fluctuation theorem for the probability distribution function. One virtue of this Langevin approach is that one does not need any counterpart of the chaotic hypothesis in the deterministic system. Under the Langevin dynamics, the system inevitably relaxes to a stationary state under time-independent driving due to the thermal noise and therefore one does not need any hypothesis to guarantee the relaxation. In this sense, the chaoticity in deterministic dynamics is replaced by the stochasticity. In Ref. [25], a general Markov process was considered. They considered the time evolution of the cumulant generating function of what they called the action functional. Then, from a symmetry of the time-evolution generator and techniques of the large deviation theory, they proved an asymptotic symmetry for the cumulant generating function, which leads to the steady-state fluctuation theorem for the action functional. For some concrete systems, they showed that, when one assumes the detailed balance condition, the action functional reduces to a quantity corresponding to the entropy production. Thus, the detailed balance in stochastic systems plays a crucial role in deriving the fluctuation theorems as the reversibility does in deterministic systems. In Ref. [26], the fluctuation theorems for thermostatted stochastic systems were also derived, where they constructed reversed trajectories corresponding to original trajectories and compared their probabilities to obtain the fluctuation theorems. Hence, the fluctuation theorems do not rely on the reversibility or the determinism of dynamics, and hence apply to stochastic systems.

In Ref. [27], the fluctuation theorem was derived on the basis of a symmetry of the space-time Gibbs measure, which is known to describe some nonequilibrium steady states in spatially extended systems.

Hence, the fluctuation theorems apply to a wide range of nonequilibrium systems regardless of thermostatting mechanisms or whether the dynamics is deterministic or stochastic.

In Ref. [28], the transient fluctuation theorem was experimentally tested in a system of a colloidal particle. They trapped a colloidal particle by an optical tweezer, and translated the trap at a constant speed. The transient fluctuation theorem does not directly apply in the frame of the laboratory since the driving is time-dependent. Nevertheless, the transient fluctuation theorem is valid in the co-moving frame. They repeatedly generated trajectories and obtained the histogram for the entropy production. In particular, they observed events that violate the second law with negative entropy productions. Moreover, the transient fluctuation theorem was confirmed by the obtained probability distribution function. Later on, in the same system, the steady-state fluctuation theorem was experimentally verified [29].

2.2 Fluctuation Theorems

In the previous section, we discuss the fluctuation theorems under time-independent driving. We here consider the fluctuation theorems that are valid even under time-dependent driving. We first consider the Jarzynski equality and the Crooks fluctuation theorem, which give relations between the nonequilibrium work distribution and the equilibrium free-energy difference. Then, we discuss the fluctuation theorems for various thermodynamic quantities. Finally, we consider a general structure of the fluctuation theorem.

2.2.1 Nonequilibrium Work Relations

2.2.1.1 Jarzynski Equality

In 1997, Jarzynski found a relation between the work and the free energy in a nonequilibrium process, which is now called the Jarzynski equality [30]. At the initial time, the state is prepared to be a canonical distribution with inverse temperature β with respect to the initial Hamiltonian. Then, one drives the system by changing parameters in the Hamiltonian for a finite time interval. By this driving, some work W is done on the system. The Jarzynski equality states that the work W satisfies

$$\langle e^{-\beta(W-\Delta F)}\rangle = 1, \tag{2.20}$$

where $\langle\cdot\rangle$ represents the ensemble average obtained from the repetition of the nonequilibrium process with the same driving, and ΔF is the equilibrium free-energy difference. By the equilibrium free-energy difference, we mean the difference between the free energy corresponding to the final Hamiltonian, which is not necessarily the nonequilibrium free energy of the final state, and the initial free energy. Remarkably, the equilibrium free-energy difference is calculated by the nonequilibrium average about the work done on the system as

$$\Delta F = -\beta^{-1}\ln\langle e^{-\beta W}\rangle. \tag{2.21}$$

Thus, we do not need to invoke quasistatic processes to evaluate the free-energy differences between two equilibrium configurations since finite-time experiments give the difference.

Moreover, the Jarzynski equality implies the second law of thermodynamics and the fluctuation-dissipation relation [30]. Since the exponential function is convex, we can apply Jensen's inequality to obtain

$$\langle e^{-\beta(W-\Delta F)}\rangle \geq e^{-\beta(\langle W\rangle-\Delta F)}, \tag{2.22}$$

which, together with the Jarzynski equality, leads to

$$\langle W\rangle \geq \Delta F. \tag{2.23}$$

This inequality indicates that the free-energy increase of the system cannot exceed the received work, which is nothing but the second law of thermodynamics. Furthermore, from Eq. (2.21), the free-energy difference can be written as a sum of the cumulants of W as

$$\Delta F = \sum_{n=1}^{\infty} (-\beta)^{n-1} \frac{c_n}{n!}, \tag{2.24}$$

where c_n is the nth cumulant of W. When the driving is slow, the distribution of work approaches the Gaussian. Therefore, only the first two terms in the sum survive, giving

$$\langle W \rangle - \Delta F = \frac{\beta c_2}{2}, \tag{2.25}$$

where $c_2 = \langle W^2 \rangle - \langle W \rangle^2$. Therefore, the dissipated work is proportional to the work fluctuation, which is one form of the fluctuation-dissipation theorem.

In addition, the Jarzynski equality gives an upper bound for the probability of the second-law violation [31]. The probability that the work W exceeds a fixed value W_0 can be evaluated as

$$\begin{aligned} \text{Prob}[W \le W_0] &= \int_{-\infty}^{W_0} dW\, P(W) \\ &\le \int_{-\infty}^{W_0} dW\, e^{\beta(W_0 - W)} P(W) \\ &= e^{\beta W_0} \langle e^{-\beta W} \rangle \\ &= e^{\beta(W_0 - \Delta F)}. \end{aligned} \tag{2.26}$$

Thus, the probability of the second-law violation is exponentially suppressed as W_0 decreases below ΔF.

The original proof of the Jarzynski equality [30] was given for the Hamiltonian system that is weakly coupled to the heat reservoir and the Nosé-Hoover system [32, 33]. Soon after, a proof for stochastic systems was given by Jarzynski himself on the basis of the master equation [34]. Later on, the Hamiltonian derivation was generalized to a system strongly coupled to a thermal reservoir [35]. Hummer and Szabo gave a variant of the Jarzynski equality, which evaluates the free-energy landscape of a biomolecule from the nonequilibrium work distribution [36].

We note that a nonequilibrium relation similar to the Jarzynski equality was found much earlier [37, 38]. In this Bochkov-Kuzovlev relation, a situation different from that of the Jarzynski equality is considered. A system is initially prepared in equilibrium with respect to a time-independent potential. Then, the system is driven under an additional time-dependent force. In this setup, the external work W^{BK} done by the time-dependent force, which does not include the work done by the time-independent potential force, satisfies $\langle e^{-\beta W^{\mathrm{BK}}} \rangle = 1$.

The Jarzynski equality was demonstrated by an experiment of a biomolecule [39]. A single RNA attached to two beads was mechanically unfolded between two configurations with different distances of the beads. By repeating the unfolding process, the free-energy difference is estimated by three methods: the quasistatic estimation $\langle W \rangle$, the linear-response estimation $\langle W \rangle - \beta c_2/2$, and the estimation by the Jarzynski equality (2.21). It was confirmed that the Jarzynski estimation gives the best fitting to the free energy.

2.2.1.2 Crooks Fluctuation Theorem

Another renowned nonequilibrium work relation is the Crooks fluctuation theorem [40, 41]. Soon after the discovery by Jarzynski, Crooks found that the comparison between the original process and its time-reversed process concisely derives the Jarzynski equality [42]. Here, by the term of the time-reversed process, we mean the process that starts from the canonical distribution corresponding to the final parameters and goes through the time-dependent driving in the time-reversed manner. The idea to consider the time reversal led Crooks to derive the following fluctuation theorem [40]:

$$\frac{P(\sigma)}{\bar{P}(-\sigma)} = e^{\sigma}, \tag{2.27}$$

where $\sigma := \beta(W - \Delta F)$ is the dissipated work in units of $k_B T$ and P ($\bar{P}$) represents the probability distribution function of the dissipated work in the original (time-reversed) process. The crucial difference from the fluctuation theorems in the previous section is that the probability in the time-reversed process $\bar{P}$ rather than P is introduced. However, when we consider driving symmetric with respect to time reversal, the time-reversed probability $\bar{P}$ becomes identical to the original probability P. Therefore, the Crooks fluctuation theorem reduces to the $P(\sigma)/P(-\sigma) = e^{\sigma}$.

The Crooks fluctuation theorem concisely reproduces the Jarzynski equality. Actually, the average in the forward process can be transformed into the average in the time-reversed process by the Crooks fluctuation theorem as

$$\begin{aligned}\langle e^{-\sigma}\rangle &= \int_{-\infty}^{\infty} d\sigma\, e^{-\sigma} P(\sigma) \\ &= \int_{-\infty}^{\infty} d\sigma\, \bar{P}(\sigma) \\ &= 1,\end{aligned} \tag{2.28}$$

where we use the normalization condition of the probability distribution function to obtain the last line. In addition, by setting $\sigma = 0$ in Eq. (2.27), we obtain

$$\sigma = 0 \Leftrightarrow P(\sigma) = \bar{P}(-\sigma). \tag{2.29}$$

Consequently, the probability distributions of work $P(W)$ and $\bar{P}(-W)$ become equal at $W = \Delta F$. Thus, the equilibrium free-energy difference can be estimated directly from the nonequilibrium work distribution functions.

In Ref. [41], the Crooks fluctuation theorem was extended to the probability of phase-space trajectories. Let $\mathscr{P}[\Gamma]$ ($\bar{\mathscr{P}}[\bar{\Gamma}]$) be the probability distribution function of an original (time-reversed) phase-space trajectory Γ ($\bar{\Gamma}$) in the original (time-reversed) process. Then, the Crooks fluctuation theorem at the trajectory level holds as

$$\frac{\mathscr{P}[\Gamma]}{\bar{\mathscr{P}}[\bar{\Gamma}]} = e^{\sigma[\Gamma]}, \tag{2.30}$$

where $\sigma[\Gamma]$ is the dissipated work corresponding to the trajectory Γ. By using this fluctuation theorem, we can derive the integral fluctuation theorem. Let $\mathscr{F}[\Gamma]$ be an arbitrary functional which is dependent on the trajectory Γ. Then, a weighted average in the original process can be calculated as

$$\begin{aligned} \langle \mathscr{F} e^{-\sigma} \rangle &= \int \mathscr{D}\Gamma \; \mathscr{F}[\Gamma] e^{-\sigma[\Gamma]} \mathscr{P}[\Gamma] \\ &= \int \mathscr{D}\Gamma \; \mathscr{F}[\Gamma] \bar{\mathscr{P}}[\bar{\Gamma}]. \end{aligned} \tag{2.31}$$

Since the Liouville measure $\mathscr{D}\Gamma$ in phase space does not change under time reversal, we obtain

$$\langle \mathscr{F} e^{-\sigma} \rangle = \int \mathscr{D}\bar{\Gamma} \; \bar{\mathscr{F}}[\bar{\Gamma}] \bar{\mathscr{P}}[\bar{\Gamma}], \tag{2.32}$$

where we define $\bar{\mathscr{F}}[\bar{\Gamma}] := \mathscr{F}[\Gamma]$ for notational convenience. Therefore, if we denote the ensemble average in the time-reversed process by $\langle \cdot \rangle^{TR}$, we obtain

$$\langle \mathscr{F} e^{-\sigma} \rangle = \langle \bar{\mathscr{F}} \rangle^{TR}. \tag{2.33}$$

When we set $\mathscr{F}[\Gamma] = 1$, this equality reduces to the Jarzynski equality. Moreover, let us set $\mathscr{F}[\Gamma] = \delta(\sigma[\Gamma] - \sigma)$. Then, since the dissipated work is odd under time reversal: $\sigma[\bar{\Gamma}] = -\sigma[\Gamma]$, we obtain $\bar{\mathscr{F}}[\bar{\Gamma}] = \delta(\sigma[\bar{\Gamma}] + \sigma)$. As a consequence, Eq. (2.33) reduces to the Crooks fluctuation theorem for the probability distribution functions of the dissipated work (2.27).

The derivations of the Crooks fluctuation theorem in Refs. [40, 41] are based on the stochastic Markov dynamics with the detailed balance condition. In a Hamiltonian system, the Crooks fluctuation theorem was derived in Ref. [43].

The Crooks fluctuation theorem was experimentally tested by repeatedly measuring the work during unfolding and refolding of an RNA molecule [44], which is a setup similar to the one to verify the Jarzynski equality [39]. The work distributions of the unfolding and refolding processes were measured. It was confirmed that these two distributions cross each other at a single point, and this point does not change by changing the speed of unfolding and refolding. According to the Crooks fluctuation theorem, this value corresponds to the free-energy difference. The obtained estimate of the free-energy difference was confirmed to agree well with a value from an independent method. Moreover, the Crooks fluctuation theorem was directly verified from the ratio of the work distribution functions.

2.2.2 *Various Fluctuation Theorems*

We here enumerate the fluctuation theorems for various entropy productions.

2.2.2.1 Fluctuation Theorems for the Excess Entropy production

In 2001, Hatano and Sasa considered the steady-state thermodynamics of the Langevin dynamics [45]. When we consider a transition between two nonequilibrium steady states in an isothermal environment, the entropy production of the reservoir can be divided into two parts as [46]

$$\Delta s^{\mathrm{r}} = \Delta s^{\mathrm{hk}} + \Delta s^{\mathrm{ex}}, \tag{2.34}$$

where the first term is the housekeeping entropy production, which is an inevitable dissipation to maintain a nonequilibrium steady state under a fixed set of external parameters, and the second term is the excess entropy production generated by the transition with a finite-speed change of the parameters. We here define the entropy production in units of k_{B}. In addition, from the instantaneous steady state $P_t^{\mathrm{ss}}(\Gamma)$, we define the nonequilibrium potential by

$$\phi_t(\Gamma) = -\ln P_t^{\mathrm{ss}}(\Gamma). \tag{2.35}$$

Then, one can show that the Hatano-Sasa relation holds:

$$\langle e^{-\Delta\phi - \Delta s^{\mathrm{ex}}} \rangle = 1, \tag{2.36}$$

where $\Delta\phi := \phi_\tau(\Gamma_\tau) - \phi_0(\Gamma_0)$ is the difference of the nonequilibrium potential, and Γ_t is the position in phase space at time t on the trajectory Γ. From Jensen's inequality, we obtain an inequality

$$\langle \Delta\phi \rangle \geq -\langle \Delta s^{\mathrm{ex}} \rangle, \tag{2.37}$$

which can be regarded as a nonequilibrium generalization of the Clausius inequality. Note that, when we consider a transition between equilibrium states, Eq. (2.36) reduces to the Jarzynski equality. This is because $\phi_t(\Gamma) = \beta(H_t(\Gamma) - F_t)$ and $\Delta s^{\mathrm{hk}} = 0$ in this situation, where H_t and F_t are the Hamiltonian and the equilibrium free energy of the system at time t, respectively.

Later on, the detailed fluctuation theorem for $\sigma^{\mathrm{HS}} := \Delta\phi + \Delta s^{\mathrm{ex}}$ was shown in Ref. [47], where the dual dynamics, in which the transition rates are modified so that the steady state remains unchanged but the steady current is reversed, is introduced. For the path probability distribution function $\bar{\mathcal{P}}^+$ in the dual and time-reversed dynamics, the detailed fluctuation theorem for the path probabilities holds:

$$\frac{\mathscr{P}[\Gamma]}{\bar{\mathscr{P}}^{+}[\bar{\Gamma}]} = e^{\sigma^{\mathrm{HS}}}. \tag{2.38}$$

As a consequence, since σ^{HS} is odd under the combined operation of dual and time reversal, we obtain the detailed fluctuation theorem:

$$\frac{P(\sigma^{\mathrm{HS}})}{\bar{P}^{+}(-\sigma^{\mathrm{HS}})} = e^{\sigma^{\mathrm{HS}}}. \tag{2.39}$$

An experimental demonstration of the Hatano-Sasa relation was given in a colloidal system [48], where the colloidal particle was dragged by an optical tweezer and its translation speed was changed from one value to another to realize a transition between nonequilibrium steady states.

2.2.2.2 Fluctuation Theorems for the Housekeeping Entropy production

The housekeeping entropy production also satisfies the integral fluctuation theorem as [49]

$$\langle e^{-\Delta s^{\mathrm{hk}}} \rangle = 1. \tag{2.40}$$

The detailed fluctuation theorem holds for the dual dynamics without time reversal as

$$\frac{\mathscr{P}[\Gamma]}{\mathscr{P}^{+}[\Gamma]} = e^{\Delta s^{\mathrm{hk}}}. \tag{2.41}$$

Since Δs^{hk} is odd under the dual operation, we also obtain

$$\frac{P(\Delta s^{\mathrm{hk}})}{P^{+}(-\Delta s^{\mathrm{hk}})} = e^{\Delta s^{\mathrm{hk}}}. \tag{2.42}$$

2.2.2.3 Fluctuation Theorems for the Total Entropy production

We here consider a process starting from an arbitrary initial state. We define the unaveraged Shannon entropy of the system by

$$s_t = -\ln P_t(\Gamma_t), \tag{2.43}$$

where P_t is the probability distribution function in phase space at time t. Then, we define the total entropy production for the system and the heat reservoir by

$$\Delta s^{\mathrm{tot}} = \Delta s + \Delta s^{\mathrm{r}}, \tag{2.44}$$

where $\Delta s := s_\tau - s_0$. For this quantity, the integral fluctuation theorem holds [50]:

$$\langle e^{-\Delta s^{\mathrm{tot}}} \rangle = 1. \tag{2.45}$$

The detailed fluctuation theorem also holds at the trajectory level as

$$\frac{\mathscr{P}[\Gamma]}{\bar{\mathscr{P}}[\bar{\Gamma}]} = e^{\Delta s^{\mathrm{tot}}}, \tag{2.46}$$

where the time-reversed process starts from the finial state of the original process. However, the detailed fluctuation theorem does not always hold for the probability distributions of the total entropy production Δs^{tot}, since Δs is not odd with respect to time reversal. This is due to the asymmetry that the final state of the time-reversed process does not necessarily coincide with the initial state of the original process, although the final state of the original process is set to be the initial state of the time-reversed process.

2.2.3 *General Structures of the Fluctuation Theorems*

In the previous section, we discuss various fluctuation theorems and see that they have common structures. We here consider their general structures. Let us first consider the detailed fluctuation theorem, since the integral fluctuation theorem can be derived from it. The detailed fluctuation theorem for path can be written as

$$\frac{\mathscr{P}[\Gamma]}{\mathscr{P}^{\mathrm{r}}[\Gamma]} = e^{\sigma[\Gamma]}, \tag{2.47}$$

where $\mathscr{P}^{\mathrm{r}}$ is the reference process and $\sigma[\Gamma]$ is the entropy production. We have seen that for physical entropy productions we can construct the corresponding reference processes. Conversely, if we specify the reference process by determining the initial state and the dynamics, the entropy production $\sigma[\Gamma]$ is formally defined. When the choice of the reference process is proper, the entropy production acquires a physical significance.

From Eq. (2.47), we obtain for an arbitrary functional $\mathscr{F}[\Gamma]$

$$\langle \mathscr{F} e^{-\sigma} \rangle = \langle \mathscr{F} \rangle^{\mathrm{r}}, \tag{2.48}$$

where $\langle \cdot \rangle^{\mathrm{r}}$ represents the average in the reference process. By setting $\mathscr{F} = 1$, we obtain the integral fluctuation theorem

$$\langle e^{-\sigma} \rangle = 1. \tag{2.49}$$

When we set $\mathscr{F}[\Gamma] = \delta(\sigma[\Gamma] - \sigma)$, the equation reduces to

$$e^{-\sigma} P(\sigma) = \langle \delta(\sigma[\Gamma] - \sigma) \rangle^{\mathrm{r}}. \tag{2.50}$$

When the entropy production has the odd parity: $\sigma^{\mathrm{r}}[\Gamma] = -\sigma[\Gamma]$, we obtain the detailed fluctuation theorem for the entropy production as

$$\frac{P(\sigma)}{P^{\mathrm{r}}(-\sigma)} = e^{\sigma}. \tag{2.51}$$

When $\sigma[\Gamma]$ has the even parity, we obtain the detailed fluctuation theorem as

$$\frac{P(\sigma)}{P^{\mathrm{r}}(\sigma)} = e^{\sigma}. \tag{2.52}$$

We have seen three choices of dynamics in the reference process: time reversal, dual and their combination. Under these dynamics, the ratios of transition probabilities are respectively related to entropy productions of the heat reservoir [51, 52]:

$$\frac{\mathscr{P}[\Gamma|\Gamma_0]}{\bar{\mathscr{P}}[\bar{\Gamma}|\bar{\Gamma}_0]} = e^{\Delta s^{\mathrm{r}}}, \tag{2.53}$$

$$\frac{\mathscr{P}[\Gamma|\Gamma_0]}{\mathscr{P}^{+}[\Gamma|\Gamma_0]} = e^{\Delta s^{\mathrm{hk}}}, \tag{2.54}$$

$$\frac{\mathscr{P}[\Gamma|\Gamma_0]}{\bar{\mathscr{P}}^{+}[\bar{\Gamma}|\bar{\Gamma}_0]} = e^{\Delta s^{\mathrm{ex}}}. \tag{2.55}$$

Therefore, the formal entropy production σ can be written as

$$\sigma = \ln \frac{P_0(\Gamma_0)}{\bar{P}_0(\bar{\Gamma}_0)} + \Delta s^{\mathrm{r}}, \tag{2.56}$$

$$\sigma = \ln \frac{P_0(\Gamma_0)}{P_0^{+}(\Gamma_0)} + \Delta s^{\mathrm{hk}}, \tag{2.57}$$

$$\sigma = \ln \frac{P_0(\Gamma_0)}{\bar{P}_0^{+}(\bar{\Gamma}_0)} + \Delta s^{\mathrm{ex}}. \tag{2.58}$$

By properly choosing the initial state of the reference processes under some assumptions, we can endow the entropy production with a physical meaning. For example, suppose that the initial state of the original process is a canonical distribution $P_0(\Gamma_0) = P_0^{\mathrm{eq}}(\Gamma_0) = e^{-\beta(H_0(\Gamma_0) - F_0)}$. Then, we set the reference initial state to be the canonical equilibrium: $\bar{P}_0(\bar{\Gamma}_0) = P_\tau^{\mathrm{eq}}(\Gamma_\tau) = e^{-\beta(H_\tau(\Gamma_\tau) - F_\tau)}$. As a consequence, the formal entropy production reduces to $\sigma = \beta(\Delta H - \Delta F) + \Delta s^{\mathrm{r}} = \beta(W - \Delta F)$. Thus, Eq. (2.47) reduces to the Crooks fluctuation theorem, which leads to the Jarzynski equality. Other choices are shown in Table 2.1. One can confirm that the bulk term Δs^{i} ($\mathrm{i} = \mathrm{r}, \mathrm{hk}, \mathrm{ex}$) is odd under the corresponding reversal operation. Hence, when the boundary term, which is the logarithmic ratio of the initial probabilities, is odd with respect to the same operation, the entropy production σ satisfies the detailed fluctuation theorem.

Table 2.1 Reference processes and the corresponding fluctuation theorems. Under an assumption and a proper choice of the reference process, the formal entropy production σ acquires a physical meaning. By construction, the detailed fluctuation theorem (DFT) for each individual path is satisfied. The detailed fluctuation theorem for the entropy production σ holds only when its value in the reference process is equal to the negative of the original value. The integral fluctuation theorem (IFT) follows from the detailed fluctuation theorem for the path

Assumption	Ref. dynamics	Ref. initial state	Entropy production σ	DFT for path	DFT for σ	IFT
Canonical initial state	Time reversal	Final canonical state: $P^{\mathrm{eq}}_\tau(\Gamma_\tau)$	Dissipative work: $\sigma^{\mathrm{dis}} = \beta(W - \Delta F)$	$\frac{\mathcal{P}[\Gamma]}{\tilde{\mathcal{P}}[\tilde{\Gamma}]} = e^{\sigma^{\mathrm{dis}}}$	$\frac{P(\sigma^{\mathrm{dis}})}{\tilde{P}(-\sigma^{\mathrm{dis}})} = e^{\sigma^{\mathrm{dis}}}$	$\langle e^{-\sigma^{\mathrm{dis}}}\rangle = 1$
–	Time reversal	Final state: $P_\tau(\Gamma_\tau)$	Total entropy production: $\Delta s^{\mathrm{tot}} = \Delta s + \Delta s^{\mathrm{r}}$	$\frac{\mathcal{P}[\Gamma]}{\tilde{\mathcal{P}}[\tilde{\Gamma}]} = e^{\Delta s^{\mathrm{tot}}}$	–	$\langle e^{-\Delta s^{\mathrm{tot}}}\rangle = 1$
Constant driving	Time reversal	Initial state: $P_0(\Gamma_\tau)$	Dissipative function: $\Omega = -\ln\left[\frac{P_0(\Gamma_\tau)}{P_0(\Gamma_0)}\right] + \Delta s^{\mathrm{r}}$	$\frac{\mathcal{P}[\Gamma]}{\mathcal{P}[\tilde{\Gamma}]} = e^{\Omega}$	$\frac{P(\Omega)}{P(-\Omega)} = e^{\Omega}$	$\langle e^{-\Omega}\rangle = 1$
Canonical initial state	Time reversal	Initial canonical state: $P^{\mathrm{eq}}_0(\Gamma_\tau)$	External work: βW^{BK}	$\frac{\mathcal{P}[\Gamma]}{\tilde{\mathcal{P}}[\tilde{\Gamma}]} = e^{\beta W^{\mathrm{BK}}}$	$\frac{P(W^{\mathrm{BK}})}{\tilde{P}(-W^{\mathrm{BK}})} = e^{\beta W^{\mathrm{BK}}}$	$\langle e^{-\beta W^{\mathrm{BK}}}\rangle = 1$
Steady initial state	Dual and time reversal	Final steady state: $P^{\mathrm{ss}}_\tau(\Gamma_\tau)$	Hatano-Sasa dissipation: $\sigma^{\mathrm{HS}} = \Delta\phi + \Delta s^{\mathrm{ex}}$	$\frac{\mathcal{P}[\Gamma]}{\tilde{\mathcal{P}}^{+}[\tilde{\Gamma}]} = e^{\sigma^{\mathrm{HS}}}$	$\frac{P(\sigma^{\mathrm{HS}})}{\tilde{P}^{+}(-\sigma^{\mathrm{HS}})} = e^{\sigma^{\mathrm{HS}}}$	$\langle e^{-\sigma^{\mathrm{HS}}}\rangle = 1$
–	Dual and time reversal	Final state: $P_\tau(\Gamma_\tau)$	Nonadiabatic entropy production: $\Delta s^{\mathrm{na}} = \Delta s + \Delta s^{\mathrm{ex}}$	$\frac{\mathcal{P}[\Gamma]}{\tilde{\mathcal{P}}^{+}[\tilde{\Gamma}]} = e^{\Delta s^{\mathrm{na}}}$	–	$\langle e^{-\Delta s^{\mathrm{na}}}\rangle = 1$
–	Dual	Initial state: $P_0(\Gamma_0)$	Housekeeping entropy production: Δs^{hk}	$\frac{\mathcal{P}[\Gamma]}{\mathcal{P}^{+}[\Gamma]} = e^{\Delta s^{\mathrm{hk}}}$	$\frac{P(\sigma^{\mathrm{hk}})}{P^{+}(-\sigma^{\mathrm{hk}})} = e^{\sigma^{\mathrm{hk}}}$	$\langle e^{-\Delta s^{\mathrm{hk}}}\rangle = 1$

2.3 Fluctuation Theorems under Nonuniform Temperature

Researches on diffusion under nonuniform temperature have a long history dating back to Landauer [53, 54], Büttiker [55] and van Kampen [56, 57]. Recently, experiments under nonuniform temperature were realized for a colloidal particle [58] and a biomolecule [59]. Accordingly, the theoretical study on nonuniform temperature attracts a renewed interest [60–63]. In 2012, Celanni et al. found that the overdamped approximation dramatically fails to describe thermodynamics under nonuniform temperature [64, 65]. To be specific, let Δs^{tot} ($\Delta s^{\mathrm{tot,od}}$) denote the total entropy production defined from the underdamped (overdamped) Langevin dynamics. Then, in the overdamped limit with the momentum relaxation time ε approaching zero, these two quantities do not coincide as

$$\Delta s^{\mathrm{tot}} \nrightarrow \Delta s^{\mathrm{tot,od}} \ (\varepsilon \to 0), \tag{2.59}$$

although the corresponding dynamics coincide in the same limit. Thus, the overdamped limit in the presence of nonuniform temperature fails to correctly evaluate thermodynamic quantities. Actually, the total entropy production in the underdamped stochastic thermodynamics can be separated as

$$\Delta s^{\mathrm{tot}} = \Delta s^{\mathrm{reg}} + \Delta s^{\mathrm{anom}}. \tag{2.60}$$

The regular part Δs^{reg} approaches the overdamped total entropy production in the overdamped limit

$$\Delta s^{\mathrm{reg}} \to \Delta s^{\mathrm{tot,od}} \ (\varepsilon \to 0). \tag{2.61}$$

Meanwhile, the anomalous part Δs^{anom} has no counterpart in the overdamped limit and originates from an asymmetry of the velocity distribution due to the temperature gradient.

The underdamped thermodynamics naturally satisfies the fluctuation theorem

$$\langle e^{-\Delta s^{\mathrm{tot}}} \rangle = 1. \tag{2.62}$$

Interestingly, the two contributions in Eq. (2.60) separately satisfy the fluctuation theorems in the overdamped limit [64, 65]

$$\langle e^{-\Delta s^{\mathrm{reg}}} \rangle = 1 \ (\varepsilon \to 0), \tag{2.63}$$

$$\langle e^{-\Delta s^{\mathrm{anom}}} \rangle = 1 \ (\varepsilon \to 0). \tag{2.64}$$

We note that this is analogous to the separation of the total entropy production into the nonadiabatic part and the housekeeping part: $\Delta s^{\mathrm{tot}} = \Delta s^{\mathrm{na}} + \Delta s^{\mathrm{hk}}$, each of which separately satisfies the integral fluctuation theorem.

References

1. Evans DJ, Cohen EGD, Morriss GP (1993) Phys Rev Lett 71:2401
2. Eckmann JP, Procaccia I (1986) Phys Rev A 34:659
3. Evans DJ, Morriss G (1990) Statistical mechanics of nonequilibrium liquids. Cambridge University Press
4. Gallavotti G, Cohen EGD (1995) Phys Rev Lett 74:2694
5. Gallavotti G, Cohen EGD (1995) J Stat Phys 80:931
6. Gallavotti G (2014) Nonequilibrium and irreversibility. Springer International Publishing
7. Sinai YG (1972) Russian Math Surveys 27:21
8. Ruelle D (1976) Am J Math 98:619
9. Bowen R, Ruelle D (1975) Inventiones Math 29:181

10. Evans DJ, Searles DJ (1994) Phys Rev E 50:1645
11. Cohen EGD, Gallavotti G (1999) J Stat Phys 96:1343
12. Searles DJ, Evans DJ (2000) J Chem Phys 112:9727
13. Gallavotti G (1996) Phys Rev Lett 77:4334
14. Green MS (1952) J Chem Phys 20:1281
15. Green MS (1954) J Chem Phys 22:398
16. Kubo R (1957) J Phys Soc Jpn 12:570
17. Kubo R, Yokota M, Nakajima S (1957) J Phys Soc Jpn 12:1203
18. Onsager L (1931) Phys Rev 37:405
19. Onsager L (1931) Phys Rev 38:2265
20. Yamada T, Kawasaki K (1967) Prog Theor Phys 38:1031
21. Evans DJ, Searles DJ (1995) Phys Rev E 52:5839
22. Searles DJ, Evans DJ (2000) J Chem Phys 113:3503
23. Evans DJ, Searles DJ, Mittag E (2001) Phys Rev E 63:051105
24. Kurchan J (1998) J Phys A Math Gen 31:3719
25. Lebowitz JL, Spohn H (1999) J Stat Phys 95:333
26. Searles DJ, Evans DJ (1999) Phys Rev E 60:159
27. Maes C (1999) J Stat Phys 95:367
28. Wang GM, Sevick EM, Mittag E, Searles DJ, Evans DJ (2002) Phys Rev Lett 89:050601
29. Wang GM, Reid JC, Carberry DM, Williams DRM, Sevick EM, Evans DJ (2005) Phys Rev E 71:046142
30. Jarzynski C (1997) Phys Rev Lett 78:2690
31. Jarzynski C (1999) J Stat Phys 96:415
32. Nosé S (1984) Mol Phys 52:255
33. Hoover WG (1985) Phys Rev A 31:1695
34. Jarzynski C (1997) Phys Rev E 56:5018
35. Jarzynski C (2004) J Stat Mech Theor Exp P09005
36. Hummer G, Szabo A (2001) Proc Natl Acad Sci USA 98:3658
37. Bochkov GN, Kuzovlev YE (1977) J Exp Theor Phys 45:125
38. Bochkov GN, Kuzovlev YE (1979) J Exp Theor Phys 49:543
39. Liphardt J, Dumont S, Smith SB, Ignacio Tinoco J, Bustamante C (2002) Science 296:1832
40. Crooks GE (1999) Phys Rev E 60:2721
41. Crooks GE (2000) Phys Rev E 61:2361
42. Crooks GE (1998) J Stat Phys 90:1481
43. Jarzynski C (2000) J Stat Phys 98:77
44. Collin D, Ritort F, Jarzynski C, Smith SB, Tinoco JI, Bustamante C (2005) Nature 437:231
45. Hatano T, Sasa S (2001) Phys Rev Lett 86:3463
46. Oono Y, Paniconi M (1998) Prog Theor Phys Suppl 130:29
47. Chernyak VY, Chertkov M, Jarzynski C (2006) J Stat Mech Theor Exp P08001
48. Trepagnier EH, Jarzynski C, Ritort F, Crooks GE, Bustamante CJ, Liphardt J (2004) Proc Natl Acad Sci USA 101:15038
49. Speck T, Seifert U (2005) J Phys A Math Gen 38:L581
50. Seifert U (2005) Phys Rev Lett 95:040602
51. Esposito M, Van den Broeck C (2010) Phys Rev Lett 104:090601
52. Seifert U (2012) Rep Prog Phys 75:126001
53. Landauer R (1975) Phys Rev A 12(2):636
54. Landauer R (1988) J Stat Phys 53:233
55. Büttiker M (1987) Z Phys B 68:161
56. van Kampen NG (1987) J Phys Chem Solids 68:135
57. van Kampen NG (1988) IBM J Res Dev 32:107
58. Duhr S, Braun D (2006) Phys Rev Lett 96:168301
59. Pedersen JN, Lüscher CJ, Marie R, Thamdrup LH, Kristensen A, Flyvbjerg H (2014) Phys Rev Lett 113:268301
60. Matsuo M, Sasa S (2000) Physica A 276:188

61. Rings D, Schachoff R, Selmke M, Cichos F, Kroy K (2010) Phys Rev Lett 105:090604
62. Polettini M (2013) Phys Rev E 87:032126
63. Sancho JM (2015) Phys Rev E 92:062110
64. Celani A, Bo S, Eichhorn R, Aurell E (2012) Phys Rev Lett 109:260603
65. Bo S, Celani A (2014) J Stat Phys 154:1325

Chapter 3
Review on Fluctuation Theorems with Absolute Irreversibility

In this chapter, we review the notion of absolute irreversibility. Although the fluctuation theorems apply to a wide range of nonequilibrium processes, they are inapplicable to free expansion [1–3]. This inapplicability is due to divergence of entropy production, which mathematically corresponds to singularities of probability measures. Therefore, by invoking measure theory, we define the notion of absolute irreversibility as singularly irreversible events, which render the conventional integral fluctuation theorems inapplicable. Then, we generalize the fluctuation theorems to situations in the presence of absolute irreversibility. This chapter is partly based on the published article [4] and the author's master thesis [5].

3.1 Apparent Breakdown of the Fluctuation Theorems

The Jarzynski equality [6] apparently breaks down for free expansion [1, 2]. For illustration, let us consider a bipartite box with two compartments with the same volume V. We enclose in one of the compartments a classical ideal gas with particle number N and let it equilibrate at temperature T. Then, we remove the partition in the middle as illustrated in Fig. 3.1a. No work is done in this process:

$$W = 0. \tag{3.1}$$

At the same time, the thermodynamic free energy decreases by the amount of

$$\Delta F = -k_B T N \ln 2. \tag{3.2}$$

As a consequence, the average in the Jarzynski equality is calculated as

$$\langle e^{-\beta(W-\Delta F)} \rangle = \frac{1}{2^N}. \tag{3.3}$$

Y. Murashita, *Fluctuation Theorems under Divergent Entropy Production and their Applications for Fundamental Problems in Statistical Physics*, Springer Theses, https://doi.org/10.1007/978-981-16-8638-2_3

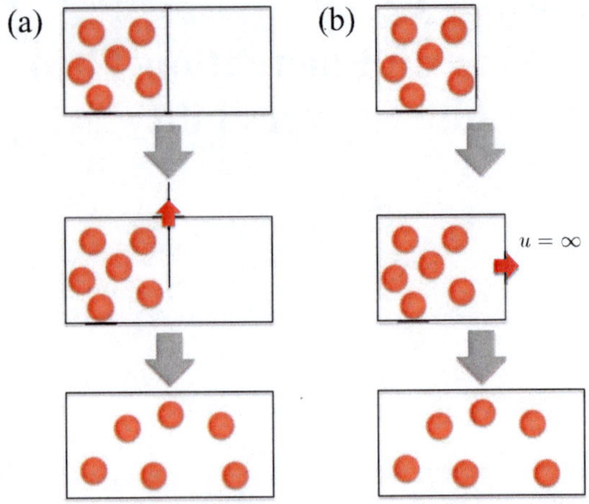

Fig. 3.1 Two implementations of free expansion by **a** removal of the partition, and **b** shift of the partition at an infinite velocity

Therefore, the Jarzynski equality is not satisfied as

$$\langle e^{-\beta(W-\Delta F)} \rangle \neq 1 \tag{3.4}$$

for the process of free expansion.

The origin of this discrepancy resides in the initial condition [3]. The Jarzynski equality assumes a canonical state corresponding to the initial Hamiltonian as the initial state. This means that the initial state should equilibrate over the entire system. In this global equilibrium, particles should be distributed equally to each compartment. However, the initial state of the free expansion equilibrates in only one of the compartments and all particles are in it; the initial state is in an equilibrium over a constrained phase space, which we refer to as a constrained equilibrium for convenience. Thus, the free expansion violates the assumption of an initial global equilibrium. Therefore, the Jarzynski equality does not break down but cannot be applied to the free expansion.

As illustrated in Fig. 3.1b, one can realize free expansion by an alternative approach of moving the partition at an infinite speed [3, 7]. Let u denote the speed of the partition. If u is much larger than the thermal velocity $v_{\rm th} \sim \sqrt{k_{\rm B}T/m}$, the work almost always vanishes. Therefore, one would again expect breakdown of the Jarzynski equality. However, rare events save the Jarzynski equality. Let us consider a particle with very high speed v ($\gtrsim u$). Then, the work done by the particle at a single collision against the partition is $-W \sim 2mvu$. Then, its contribution to the Jarzynski average is approximately

$$e^{2\beta mvu} e^{-\beta mv^2/2}, \tag{3.5}$$

whose peak is achieved when $v = 2u$. Therefore, atypical events with extremely large velocity ($v \sim 2u$) and tiny probability ($\sim e^{-2\beta mu^2}$) are indispensable for the convergence of the Jarzynski equality. This fact is analytically and numerically demonstrated in Ref. [7]. Thus, in this realization of the free expansion, the Jarzynski equality is valid, although its convergence becomes terribly slow as u goes to infinity. Again, the assumption of a canonical distribution as the initial state plays a key role in that contributions by its far tail of the Maxwellian velocity distribution are crucial. As a consequence, before we sample these extremely rare events, the Jarzynski average in the free expansion by infinitely fast wall motion coincides with that in free expansion by wall removal, and the Jarzynski equality apparently breaks down.

3.2 Absolute Irreversibility

In the previous section, we see that the Jarzynski equality is inapplicable to free expansion by wall removal. (We hereafter consider only the free expansion of this type and simply refer to it as free expansion.) This is because the initial state is not a global but only a constrained equilibrium state. However, this consequence is unsatisfactory in that the fluctuation theorem does not apply to such a fundamental process of thermodynamics. Hence, an extension of the fluctuation theorem to this kind of situations is needed. To this aim, we physically and mathematically characterize the inapplicability of the fluctuation theorem. As a result, we introduce a notion of absolute irreversibility, which encompasses situations where the integral fluctuation theorem is inapplicable.

3.2.1 Divergent Entropy Production in Free Expansion

For illustration, we here consider free expansion of a single-particle gas. The forward process of the free expansion is wall removal as illustrated in Fig. 3.2a. To formulate the fluctuation theorem, it is convenient to consider the backward process (see Fig. 3.2b). The backward process starts from the equilibrium state over the entire box at temperature T. The partition is inserted in the place where it was in the initial state of the forward process. Then, the particle ends up in either the left or the right box.

Let us consider a set of virtual paths $\{\Gamma_{\mathrm{R}}\}$ starting from the right box in the forward process. By construction of the forward process, their probability vanishes: $\mathcal{P}[\Gamma_{\mathrm{R}}] = 0$. On the other hand, these paths have nonzero probabilities in the backward process: $\bar{\mathcal{P}}[\bar{\Gamma}_{\mathrm{R}}] \neq 0$. Thus, in the process of free expansion, the following condition is satisfied:

$$^{\exists}\Gamma,\ \mathcal{P}[\Gamma] = 0\ \&\ \bar{\mathcal{P}}[\bar{\Gamma}] \neq 0. \tag{3.6}$$

Therefore, in the context of the detailed fluctuation theorem, the probability ratio diverges:

$$\frac{\bar{\mathcal{P}}[\bar{\Gamma}]}{\mathcal{P}[\Gamma]} = \infty. \tag{3.7}$$

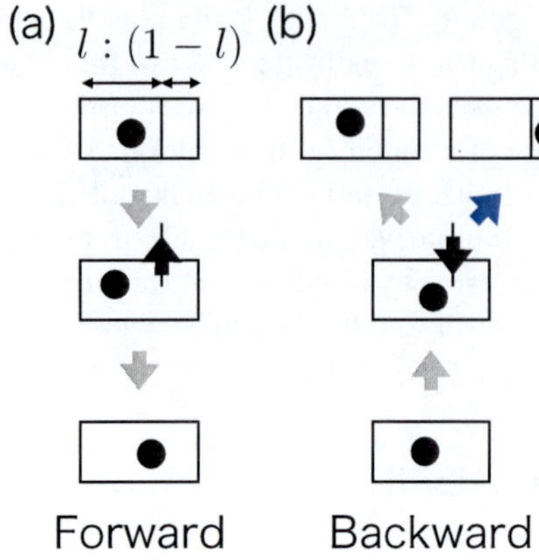

Fig. 3.2 Free expansion of a single-particle gas by wall removal and its time reversal by wall insertion. **a** Forward process of free expansion. The box is divided according to the ratio of l to $1-l$ by a partition. A single-particle gas is initially enclosed in the left compartment. When we remove the wall, the gas freely expands to the entire box. **b** Backward process of free expansion, i.e., wall insertion. Initially, a single-particle gas is in a thermal equilibrium over the entire box. We insert the partition, and then the gas ends up in either the left or the right compartment. The events with the particle ending up in the right box has no counterparts in the forward process, resulting in divergent entropy production as indicated by the detailed fluctuation theorem. Reprinted figure with permission from Ref. [4]. Copyright 2014 by the American Physical Society

As a consequence, the entropy production $\sigma := \beta(W - \Delta F)$ negatively diverges as

$$\sigma = -\ln \frac{\bar{\mathscr{P}}[\bar{\Gamma}]}{\mathscr{P}[\Gamma]} = -\infty. \tag{3.8}$$

In the presence of this divergence, the Jarzynski average can be calculated as

$$\begin{aligned}
\langle e^{-\sigma} \rangle &= \int_{\mathscr{P}[\Gamma]\neq 0} e^{-\sigma} \mathscr{P}[\Gamma] \mathscr{D}\Gamma \\
&= \int_{\mathscr{P}[\Gamma]\neq 0} \frac{\bar{\mathscr{P}}[\bar{\Gamma}]}{\mathscr{P}[\Gamma]} \mathscr{P}[\Gamma] \mathscr{D}\Gamma \\
&= \int_{\mathscr{P}[\Gamma]\neq 0} \bar{\mathscr{P}}[\bar{\Gamma}] \mathscr{D}\Gamma \\
&= 1 - \int_{\mathscr{P}[\Gamma]=0} \bar{\mathscr{P}}[\bar{\Gamma}] \mathscr{D}\bar{\Gamma} < 1,
\end{aligned} \tag{3.9}$$

where we use the invariance of phase-space volume under time reversal: $\mathscr{D}\Gamma = \mathscr{D}\bar{\Gamma}$. Hence, these singular behaviors of the probability ratio and the entropy production lead to the inapplicability of the fluctuation theorem.

In ordinary situations, the probabilities satisfy

$$^{\forall}\Gamma,\ \mathscr{P}[\Gamma]=0 \Rightarrow \bar{\mathscr{P}}[\bar{\Gamma}]=0. \tag{3.10}$$

Therefore, the probability ratio is always well-defined and the entropy production is finite (or at worst positively divergent). Thus, in ordinary irreversible situations, if the backward probability is nonzero, so does the forward probability. In this sense, these paths are stochastically reversible, although they are not thermodynamically so. In contrast, if the condition (3.6), which is the negation of the condition (3.10), is satisfied, the paths are not only thermodynamically irreversible but also stochastically irreversible. Thus, we shall call these paths *absolutely irreversible*. Moreover, we refer to the processes that involve absolutely irreversible paths as *absolutely irreversible processes*.

3.2.2 Absolute Irreversibility as the Singularity of Measure

In the previous section, we see that the ill-defined probability ratio causes the inapplicability of the Jarzynski equality. Since measure theory in mathematics gives us a criterion of whether we can take the ratio, we here formulate absolute irreversibility by using measure theory.

Let $\mathscr{M}[\cdot]$ denote the path probability measure in the original process. Then, the probability for a region E of the path space $\mathfrak{P}$ is written as $\mathscr{M}[E]$. When K is an infinitesimal region as $E=\mathscr{D}\Gamma$, the probability for this region is written as $\mathscr{M}[\mathscr{D}\Gamma]$. It is ordinarily equal to $\mathscr{P}[\Gamma]\mathscr{D}\Gamma$, where $\mathscr{P}[\Gamma]$ is the probability density of forward paths and assumed to be well-defined.

To formulate the fluctuation theorem, we introduce a reference process, such as the time-reversed process and the dual process, and denote its probability measure by $\mathscr{M}^{\mathrm{r}}[\cdot]$. When the probability measures satisfy for all region E of the path space

$$\mathscr{M}[E]=0 \Rightarrow \mathscr{M}^{\mathrm{r}}[E]=0, \tag{3.11}$$

$\mathscr{M}^{\mathrm{r}}$ is said to be absolutely continuous with respect to $\mathscr{M}$. Under absolute continuity, we can take ratio between them as

$$\mathscr{M}^{\mathrm{r}}[\mathscr{D}\Gamma]=\left.\frac{\mathscr{D}\mathscr{M}^{\mathrm{r}}}{\mathscr{D}\mathscr{M}}\right|_{\Gamma}\mathscr{M}[\mathscr{D}\Gamma]. \tag{3.12}$$

This formation of the probability ratio is guaranteed by the Radon-Nykodým theorem in measure theory. For detailed explanation of mathematics, see mathematical textbooks [8, 9] or the author's master thesis [5]. Due to the detailed fluctuation theorem, this probability ratio corresponds to the entropy production as

$$\left.\frac{\mathscr{D}\mathscr{M}^{\mathrm{r}}}{\mathscr{D}\mathscr{M}}\right|_{\Gamma}=e^{-\sigma[\Gamma]}. \tag{3.13}$$

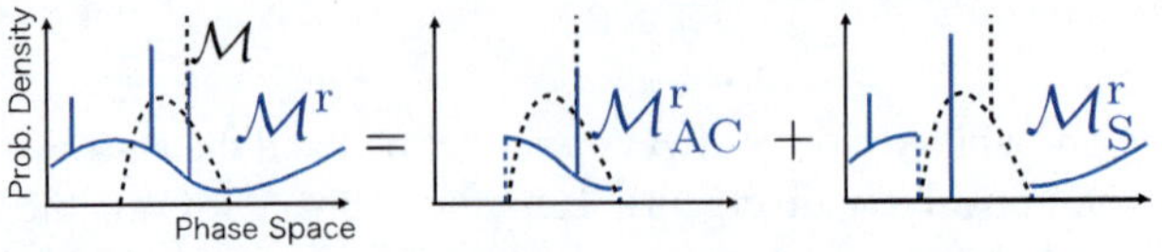

Fig. 3.3 Schematic illustration of the Lebesgue decomposition. The abscissa represents extension of phase space, and the ordinate indicates the probability density. Vertical lines schematically represent δ-function-type localization of the probability density. The reference probability measure $\mathscr{M}^{\mathrm{r}}$ (blue solid curve) is uniquely decomposed into two parts with respect to the original probability measure $\mathscr{M}$ (black dashed curve). In the former part $\mathscr{M}^{\mathrm{r}}_{\mathrm{AC}}$, the probability ratio is well-defined and the entropy production is finite. In the latter part $\mathscr{M}^{\mathrm{r}}_{\mathrm{S}}$, the probability ratio is ill-defined. Therefore, the latter corresponds to absolute irreversibility, which renders the fluctuation theorem inapplicable. Reprinted figure with permission from Ref. [4]. Copyright 2014 by the American Physical Society

Therefore, the integral fluctuation theorem is verified as

$$\begin{aligned} \langle e^{-\sigma} \rangle &:= \int e^{-\sigma[\Gamma]} \mathscr{M}[\mathscr{D}\Gamma] \\ &= \int \left. \frac{\mathscr{D}\mathscr{M}^{\mathrm{r}}}{\mathscr{D}\mathscr{M}} \right|_{\Gamma} \mathscr{M}[\mathscr{D}\Gamma] \\ &= \int \mathscr{M}^{\mathrm{r}}[\mathscr{D}\Gamma] \\ &= 1, \end{aligned} \tag{3.14}$$

where we use the normalization of the probability measure to obtain the last equality. Thus, the integral fluctuation theorem can be shown if the reference probability measure $\mathscr{M}^{\mathrm{r}}$ is absolutely continuous with respect to the original probability measure $\mathscr{M}$.

Then, how can we extend the integral fluctuation theorem when the condition of absolute continuity is violated? According to the Lebesgue decomposition theorem [5, 8, 9], $\mathscr{M}^{\mathrm{r}}$ can be uniquely decomposed into two parts as

$$\mathscr{M}^{\mathrm{r}} = \mathscr{M}^{\mathrm{r}}_{\mathrm{AC}} + \mathscr{M}^{\mathrm{r}}_{\mathrm{S}}, \tag{3.15}$$

where $\mathscr{M}^{\mathrm{r}}_{\mathrm{AC}}$ and $\mathscr{M}^{\mathrm{r}}_{\mathrm{S}}$ are absolutely continuous and singular with respect to $\mathscr{M}$ as illustrated in Fig. 3.3. Mathematically speaking, $\mathscr{M}^{\mathrm{r}}_{\mathrm{S}}$ is said to be singular with respect to $\mathscr{M}$ if there are sets K_1 and K_2 that satisfy $\mathfrak{P} = K_1 \cup K_2$, $\emptyset = K_1 \cap K_2$ and $\mathscr{M}[K_1] = \mathscr{M}^{\mathrm{r}}_{\mathrm{S}}[K_2] = 0$, where $\mathfrak{P}$ is the entire path space. Due to the Radon-Nikodým theorem, the probability ratio is well-defined for the former part. Physically, this part corresponds to ordinary irreversible paths with finite entropy production. For the latter part, we cannot take the ratio of the probabilities. Therefore, it corresponds to the absolutely irreversible part. If $\mathscr{M}^{\mathrm{r}}_{\mathrm{S}}$ exists, we call the process absolutely irreversible. In addition, a path is called absolutely irreversible if it belongs to the support of $\mathscr{M}^{\mathrm{r}}_{\mathrm{S}}$.

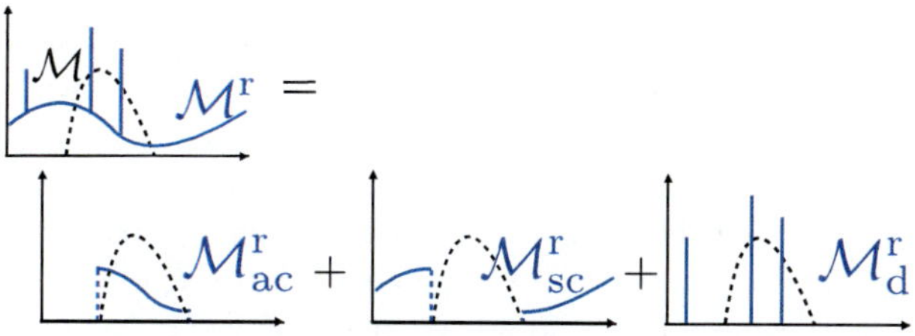

Fig. 3.4 Schematic illustration of the stronger version of the Lebesgue decomposition. The reference probability measure $\mathcal{M}^{\mathrm{r}}$ (blue curve) is decomposed into three parts with respect to the original probability measure $\mathcal{M}$ (black dashed curve). The first part is absolutely continuous with the well-defined probability ratio. In the second part, the measure is singular continuous and the probability ratio diverges due to the vanishing denominator. The third part is discrete and constitutes of δ-function-type localizations, where the probability ratio diverges since the numerator diverges as $\delta(0) = \infty$. Reprinted figure with permission from Ref. [4]. Copyright 2014 by the American Physical Society

Table 3.1 Classification of irreversibility. Ordinary irreversibility is mathematically characterized by absolute continuity of the reference probability measure. Then, the Radon-Nikodým theorem guarantees that the probability ratio is finite. Absolute irreversibility can be classified into two classes. The first class is mathematically singular continuous, and the probability ratio diverges since the forward probability density vanishes. In the second class characterized by discrete measure, the probability ratio diverges since the reference probability density itself diverges

Class of irreversibility	Ordinary	Absolute I	Absolute II
Mathematical classification	Absolutely continuous	Singular continuous	Discrete
Probability ratio	$\dfrac{\mathcal{P}^{\mathrm{r}}[\Gamma]}{\mathcal{P}[\Gamma]} = \text{finite}$	$\dfrac{\mathcal{P}^{\mathrm{r}}[\Gamma]}{0} = \infty$	$\dfrac{\delta(0)}{\mathcal{P}[\Gamma]} = \infty$

To give a more detailed picture, we consider a stronger version of the Lebesgue decomposition theorem. As illustrated in Fig. 3.4, if the forward probability measure $\mathcal{M}$ can be written in terms of the probability density as $\mathcal{M}[\mathcal{D}\Gamma] = \mathcal{P}[\Gamma]\mathcal{D}\Gamma$, the reference probability measure $\mathcal{M}^{\mathrm{r}}$ can uniquely be decomposed into three parts as

$$\mathcal{M}^{\mathrm{r}} = \mathcal{M}^{\mathrm{r}}_{\mathrm{ac}} + \mathcal{M}^{\mathrm{r}}_{\mathrm{sc}} + \mathcal{M}^{\mathrm{r}}_{\mathrm{d}}, \tag{3.16}$$

where $\mathcal{M}^{\mathrm{r}}_{\mathrm{ac}}$ is absolutely continuous with respect to $\mathcal{M}$ and corresponds to the ordinary irreversible part. The second part $\mathcal{M}^{\mathrm{r}}_{\mathrm{sc}}$ is singular with respect to $\mathcal{M}$ and continuous in the sense that it does not have any δ-function-type localization. For this part, the probability ratio is ill-defined since we divide a nonzero probability by zero as in free expansion. The third part $\mathcal{M}^{\mathrm{r}}_{\mathrm{d}}$ is discrete, namely, an ensemble of δ-function-type probability localizations. For this part, the probability ratio is ill-defined since the numerator is divergent. In this way, absolute irreversibility can be classified into the two classes as summarized in Table 3.1.

In this section, we see that the inapplicability of the fluctuation theorem is rooted physically in the divergence of the entropy production and mathematically in the singularity of probability measure. Using the Lebesgue decomposition, we define absolutely irreversible paths with mathematical rigor. Absolute irreversibility can be further classified into two classes corresponding to the singular continuous measure and the discrete measure. The former correspond to situations where we have non-vanishing reference paths that have no counterparts in the original process as in free expansion. The latter corresponds to the situation where the reference probability has δ-function-type probability localizations. We note that inapplicability of the fluctuation theorem to the latter situation had not been recognized before our study [4, 5]. The measure-theoretic formulation, together with the Lebesgue decomposition, enables us to identify this situation. An explicit example is given in the next section.

3.3 Fluctuation Theorems with Absolute Irreversibility

In the previous chapter, we mathematically formulate absolute irreversibility. By using this formulation, we here generalize the fluctuation theorem in the presence of absolute irreversibility. Then, we give some illustrative examples to demonstrate the validity of the fluctuation theorem with absolute irreversibility.

3.3.1 *Derivation*

First of all, we can take the ratio for the ordinary irreversible part as

$$\mathcal{M}^{\mathrm{r}}_{\mathrm{AC}}[\mathcal{D}\Gamma] = \left.\frac{\mathcal{D}\mathcal{M}^{\mathrm{r}}_{\mathrm{AC}}}{\mathcal{D}\mathcal{M}}\right|_{\Gamma} \mathcal{M}[\mathcal{D}\Gamma]. \tag{3.17}$$

To this well-defined probability ratio, the detailed fluctuation theorem applies as

$$\left.\frac{\mathcal{D}\mathcal{M}^{\mathrm{r}}_{\mathrm{AC}}}{\mathcal{D}\mathcal{M}}\right|_{\Gamma} = e^{-\sigma[\Gamma]}. \tag{3.18}$$

If $\mathcal{M}$ and $\mathcal{M}^{\mathrm{r}}_{\mathrm{AC}}$ can be written in terms of probability densities as $\mathcal{M}[\mathcal{D}\Gamma] = \mathcal{P}[\Gamma]\mathcal{D}\Gamma$ and $\mathcal{M}^{\mathrm{r}}_{\mathrm{AC}}[\mathcal{D}\Gamma] = \mathcal{P}^{\mathrm{r}}[\Gamma]\mathcal{D}\Gamma$, Eq. (3.18) reduces to the standard form as $\mathcal{P}^{\mathrm{r}}[\Gamma]/\mathcal{P}[\Gamma] = e^{-\sigma[\Gamma]}$. The integral of the absolutely continuous measure can be evaluated as

$$\begin{aligned}\int \mathcal{M}^{\mathrm{r}}_{\mathrm{AC}}[\mathcal{D}\Gamma] &= \int \left.\frac{\mathcal{D}\mathcal{M}^{\mathrm{r}}_{\mathrm{AC}}}{\mathcal{D}\mathcal{M}}\right|_{\Gamma} \mathcal{M}[\mathcal{D}\Gamma] \\ &= \int e^{-\sigma[\Gamma]}\mathcal{M}[\mathcal{D}\Gamma] \\ &= \langle e^{-\sigma}\rangle.\end{aligned} \tag{3.19}$$

On the other hand, since the reference probability is normalized, we obtain

$$\int \mathcal{M}^{\mathrm{r}}_{\mathrm{AC}}[\mathcal{D}\Gamma] = 1 - \int \mathcal{M}^{\mathrm{r}}_{\mathrm{S}}[\mathcal{D}\Gamma]. \tag{3.20}$$

Therefore, the integral fluctuation theorem is derived as

$$\langle e^{-\sigma} \rangle = 1 - \lambda, \tag{3.21}$$

where we define the degree of absolute irreversibility as

$$\lambda = \int \mathcal{M}^{\mathrm{r}}_{\mathrm{S}}[\mathcal{D}\Gamma]. \tag{3.22}$$

We note that λ is uniquely determined by the Lebesgue decomposition.

If the assumption of the stronger version of the Lebesgue decomposition (the absolute continuity of $\mathcal{M}$ with respect to the Lebesgue measure $\mathcal{D}\Gamma$) is satisfied, λ can be separated into two parts as

$$\lambda = \lambda_{\mathrm{sc}} + \lambda_{\mathrm{d}}, \tag{3.23}$$

where we define

$$\lambda_{\mathrm{sc}} = \int \mathcal{M}^{\mathrm{r}}_{\mathrm{sc}}[\mathcal{D}\Gamma],\ \lambda_{\mathrm{d}} = \int \mathcal{M}^{\mathrm{r}}_{\mathrm{d}}[\mathcal{D}\Gamma]. \tag{3.24}$$

The former probability λ_{sc} can be calculated by accumulating the probabilities of those reference paths that do not have the corresponding original paths. On the other hand, the latter probability λ_{d} can be calculated as the sum of probability of δ-function-type localizations.

3.3.2 Physical Implications

By applying Jensen's inequality $\langle e^{-\sigma} \rangle \geq e^{-\langle \sigma \rangle}$ to the generalized fluctuation theorem, we obtain

$$\langle \sigma \rangle \geq -\ln(1 - \lambda) \geq 0. \tag{3.25}$$

Therefore, even in the presence of absolute irreversibility, the second law of thermodynamics holds. Moreover, when we have a nonzero degree of absolute irreversibility ($\lambda > 0$), the averaged entropy production should be strictly positive. In this sense, Eq. (3.25) imposes a stronger restriction on the entropy production than the ordinary second law of thermodynamics.

The reference probability can be chosen at our disposal. Suppose that the initial state starts from a constrained equilibrium state. Then, if we set the reference initial state to be the global equilibrium state and the reference dynamics to be the time-reversed dynamics, the entropy production is given by

$$\sigma = \beta(W - \Delta F). \tag{3.26}$$

Therefore, we obtain the Jarzynski equality generalized to situations starting from a constrained equilibrium as

$$\langle e^{-\beta(W-\Delta F)} \rangle = 1 - \lambda. \tag{3.27}$$

The corresponding inequality

$$-\langle W \rangle \leq -\Delta F + \ln(1 - \lambda) \tag{3.28}$$

indicates that the averaged extractable work ($-\langle W \rangle$) decreases in the presence of absolute irreversibility since $\ln(1 - \lambda) \leq 0$.

If we set the reference initial state to be the final state of the forward process and the reference dynamics to be the time-reversed dynamics, the entropy production reduces to the total entropy production $\sigma = \Delta s^{\mathrm{tot}}$, which is the sum of the Shannon entropy production of the system and the thermodynamic entropy production of the heat bath. Hence, we obtain the generalized integral fluctuation theorem as

$$\langle e^{-\Delta s^{\mathrm{tot}}} \rangle = 1 - \lambda. \tag{3.29}$$

We can obtain fluctuation theorems with other types of entropy production by choosing other reference probabilities [10, 11].

3.3.3 Examples

We here verify the fluctuation theorem with absolute irreversibility in some illustrative examples.

3.3.3.1 Free Expansion

First of all, we consider the free expansion of a single-particle gas (see Fig. 3.2). In the forward process, the gas is initially in a constrained equilibrium state in the left compartment. We assume that the box is divided into two parts having the volume ratio $l : (1 - l)$ with $0 < l < 1$. Then, the partition is removed and the gas expands to the entire box. The Jarzynski equality generalized in the presence of absolute irreversibility is written as

$$\langle e^{-\beta(W-\Delta F)} \rangle = 1 - \lambda_{\mathrm{sc}} - \lambda_{\mathrm{d}}. \tag{3.30}$$

In this process, no work is done

$$W = 0. \tag{3.31}$$

However, the free energy decreases as

$$\Delta F = k_B T \ln l \ (< 0). \tag{3.32}$$

Therefore, the Jarzynski average is

$$\langle e^{-\beta(W-\Delta F)} \rangle = l. \tag{3.33}$$

To evaluate the degrees of absolute irreversibility, we consider the backward process. The backward process starts from the equilibrium state over the entire box. Then, after the insertion of the partition, the gas is in the left box with probability l and in the right box with probability $1 - l$. The events ending in the right box are absolutely irreversible since they have no corresponding forward events; the forward events starting from the right box are absent. Therefore, the events ending in the right box in the backward process contribute to the singular continuous probability and therefore we obtain

$$\lambda_{\text{sc}} = 1 - l. \tag{3.34}$$

Since no single backward path has positive probability, the discrete probability vanishes:

$$\lambda_{\text{d}} = 0. \tag{3.35}$$

Hence, the modified fluctuation theorem (3.30) is verified.

3.3.3.2 Langevin Dynamics Starting from a Constrained Equilibrium

Next, we consider a more complicated example with singular continuous absolute irreversibility. We consider a colloidal particle with the overdamped Langevin dynamics

$$\dot{x}_t = -\mu \partial_x U_t(x_t) + \zeta_t, \tag{3.36}$$

where x_t is the position of the particle, and μ, U and ζ are the mobility, the potential energy, and the thermal random force, respectively. The random force satisfies $\langle \zeta_t \zeta_s \rangle = 2D\delta(t-s)$, where D is the diffusion constant. In this system, the Einstein relation $D = \mu k_B T$ is satisfied.

The particle is assumed to be confined in a one-dimensional ring with length L. As illustrated in Fig. 3.5a, the potential of the system is set to be an array of n harmonic potentials as

$$U_t(x) = \begin{cases} \frac{1}{2}k_t x^2 & (-a < x \le a); \\ U_t(x - 2a\left[\frac{x+a}{2a}\right]) & (\text{otherwise}), \end{cases} \tag{3.37}$$

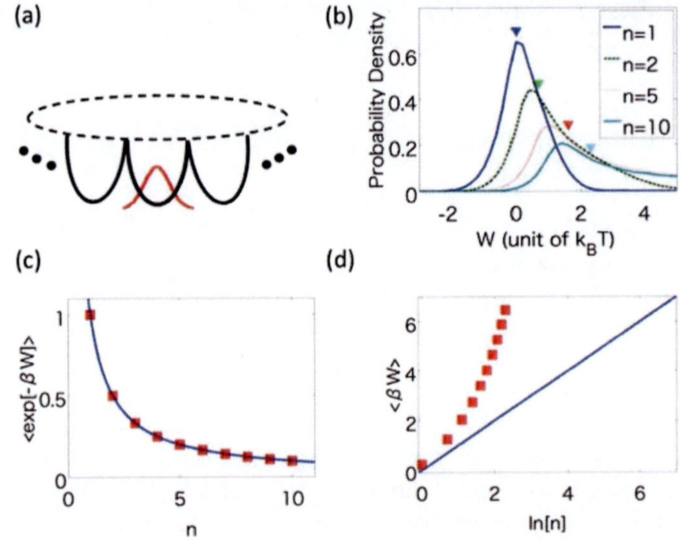

Fig. 3.5 Langevin dynamics starting from a constrained equilibrium state. **a** Schematic illustration of the potential consisting of n harmonic potentials subject to the periodic boundary condition. **b** Probability density of work done on the system for some values of n. The triangular marks indicate the values of $W = k_B T \ln n$. **c** Verification of the fluctuation theorem with absolute irreversibility. The red squares represent the numerically obtained average $\langle e^{-\beta W} \rangle$. The blue curve indicates $1/n$. **d** Verification of the second-law-like inequality. The red squares represent the averaged work $\langle \beta W \rangle$. The blue line represents the minimum dissipation $k_B T \ln n$. The parameters are set as follows: diffusion constant $D = 10^{-13}$ m^2/s; temperature $T = 300$ K; duration of the process $\tau = 10$ sec; half width of the potential $a = 10^{-6}$ m; the initial stiffness of the potential is set to satisfy $Ka^2/2 = 5k_B T$. To obtain the averages, the nonequilibrium process is repeated 10^6 times for each n. Reprinted figure with permission from Ref. [4]. Copyright 2014 by the American Physical Society

where $[\cdot]$ represents Gauss' floor function, and the position is evaluated by the modulo of $L = 2na$. The initial distribution of the forward process is set to be a constrained equilibrium within one well with its tail truncated outside the well. At the initial time $t = 0$, the stiffness of the potential is $k_{t=0} = K$. The stiffness is decreased to zero at a constant rate over time $\tau/2$ and then increased to $n^2 K$ over the next time interval $\tau/2$ as

$$k_t = \begin{cases} \left(1 - \frac{2t}{\tau}\right) K & (0 \le t \le \tau/2); \\ \left(\frac{2t}{\tau} - 1\right) n^2 K & (\tau/2 < t \le \tau). \end{cases} \tag{3.38}$$

Over this process, the work done on the system is calculated as

$$W[x] = \int_0^\tau dt\, \partial_t U_t(x_t). \tag{3.39}$$

To consider the fluctuation theorem, we set the initial state of the backward process to be the global equilibrium state and the backward dynamics to the time-reversed dynamics. Then, the fluctuation theorem reads

$$\langle e^{-\beta(W-\Delta F)} \rangle = 1 - \lambda_{\rm sc} - \lambda_{\rm d}. \tag{3.40}$$

Due to the symmetry of the initial state and the dynamics of the backward process, a backward path ends in a certain well with probability $1/n$. Since the backward paths that terminate outside the initial well do not have counterparts in the forward process, their probability contributes to the degree of singular continuous absolute irreversibility. Therefore, we obtain $\lambda_{\rm sc} = (n-1)/n$. On the other hand, since there is no single path with positive probability, we obtain $\lambda_{\rm d} = 0$. Hence, the fluctuation theorem with absolute irreversibility reads

$$\langle e^{-\beta(W-\Delta F)} \rangle = \frac{1}{n}. \tag{3.41}$$

The corresponding second-law-like inequality is

$$\langle W \rangle \geq \Delta F + k_{\rm B} T \ln n. \tag{3.42}$$

If we assume that K is sufficiently large, ΔF vanishes. In Fig. 3.5b, the probability densities of work for some values of n's are shown. Using these probability densities, we calculate the averages in the fluctuation theorem as shown in Fig. 3.5c, which are in good agreement with the theoretical value $1/n$. The averaged works are also calculated as shown in Fig. 3.5d, which are verified to be larger than the minimum dissipation indicated by Eq. (3.42). We note that this process can be regarded as information erasure of an n-digit memory. In this sense, Eq. (3.42) is a generalization of the Landauer principle [12] for a symmetric bit memory to the symmetric n-digit memory.

3.3.3.3 Langevin Dynamics with a Trap

Finally, we consider an example with discrete absolute irreversibility. The equation of motion is again given by Eq. (3.36). The potential is chosen to be a harmonic potential

$$U_t(x) = \frac{1}{2} k_t x^2. \tag{3.43}$$

We assume that there is a trapping point at $x = x_{\rm c}$ as illustrated in Fig. 3.6a. When the particle reaches this point, it is trapped with unit probability and cannot escape from the trap forever. The initial state is set to be the equilibrium state with respect to the harmonic potential (3.43). The control protocol is set to be

$$k_t = \begin{cases} \left(1 - \frac{2t}{\tau}\right) K \ (0 \leq t \leq \tau/2); \\ \left(\frac{2t}{\tau} - 1\right) K \ (\tau/2 < t \leq \tau). \end{cases} \tag{3.44}$$

Suppose that the initial state of the backward process is set to be the final state of the forward process and that the backward dynamics is set to be the time-reversed dynamics. Then, the fluctuation theorem is

$$\langle e^{-\Delta s^{\rm tot}} \rangle = 1 - \lambda_{\rm sc} - \lambda_{\rm d}, \tag{3.45}$$

where the total entropy production can be calculated as the sum of the Shannon entropy production

$$\Delta s = -\ln p_\tau(x_\tau) + \ln p_0(x_0) \tag{3.46}$$

and the thermodynamic entropy production in the heat bath

$$\beta Q[x] = -\beta \int_0^\tau dt\ \dot{x}_t \partial_x U_t(x_t). \tag{3.47}$$

Let $p_{\rm trap}$ and $p^\dagger_{\rm trap}$ denote the trapping probabilities in the final states in the forward and backward processes, respectively. In the backward process, when we start from the trapped state, the state remains trapped. Therefore, this single path has positive probability with $p_{\rm trap}$. Thus, we obtain $\lambda_{\rm d} = p_{\rm trap}$. Moreover, we have backward paths that start outside the trap and then fall into the trap. These paths do not have the corresponding forward paths since no paths can pop out of the trap in the forward process. The probability for these paths is $\lambda_{\rm sc} = p^\dagger_{\rm trap} - p_{\rm trap}$. As a result, the fluctuation theorem is

$$\langle e^{-\Delta s^{\rm tot}} \rangle = 1 - p^\dagger_{\rm trap}. \tag{3.48}$$

The corresponding inequality is

$$\langle \Delta s^{\rm tot} \rangle \geq -\ln(1 - p^\dagger_{\rm trap}). \tag{3.49}$$

To numerically demonstrate the validity of the fluctuation theorem (3.48), we obtain the probability densities for some values of τ's as shown in Fig. 3.6b. Then, we plot the average $\langle e^{-\Delta s^{\rm tot}} \rangle$ against the backward trapping probability $p^\dagger_{\rm trap}$ as shown in Fig. 3.6c, and confirm that the fluctuation theorem (3.48) is satisfied. The inequality (3.49) is automatically satisfied since the averaged entropy positively diverges due to the paths that fall into the trap in the forward process with positively divergent entropy production.

3.3.3.4 Second-Order Phase Transition

Absolute irreversibility naturally emerges in quench dynamics across phase transition [13]. As a paradigmatic example, let us consider the Ising model with ferromagnetic interactions. We prepare the initial state below the critical temperature. Then, due to the spontaneous symmetry breaking, the initial state is constrained in

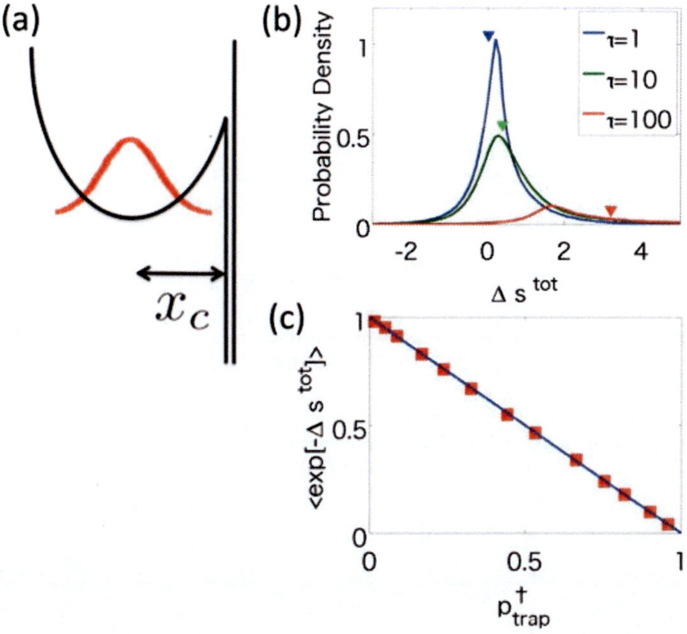

Fig. 3.6 Langevin dynamics with a trap. **a** Schematic illustration of the potential. A trapping center is located at $x = x_c$. **b** The probability densities of the total entropy production for some values of τ's. The triangular markers indicate the values of $-\ln(1 - p^\dagger_{\rm trap})$. **c** Verification of the fluctuation theorem with absolute irreversibility. The averages $\langle e^{-\Delta s^{\rm tot}}\rangle$ are plotted as the red markers against the backward trapping probability $p^\dagger_{\rm trap}$, and agree well with $1 - p^\dagger_{\rm trap}$ as indicated by the blue line. The diffusion constant D and the temperature T are the same as in the caption of Fig. 3.5. The location of the trap is $x_c = 10^{-6}$ m. The initial stiffness K is set to satisfy $Ka^2/2 = 10k_BT$. The duration of the process τ is varied from 1 sec to 100 sec to change the trapping probability. The nonequilibrium process is repeated 10^6 times for each τ. Reprinted figure with permission from Ref. [4]. Copyright 2014 by the American Physical Society

an equilibrium with positive or negative magnetization. Without loss of generality, we assume that the initial state has positive magnetization. We quench the system by decreasing the coupling constant across the critical value. Then, the system is in the disordered phase at the final time. In the time-reversed process, we prepare the system to be at equilibrium over the entire phase space. As a result, after we increase the coupling constant in the time-reversed manner, the state ends up with positive or negative spontaneous magnetization with the equal probability. Since the forward process starts with positive magnetization, the latter case contributes to absolute irreversibility. Hence, the integral fluctuation theorem reduces to $\langle e^{-\beta(W-\Delta F)}\rangle = 1/2$. In Ref. [13], effects of the slow convergence of the Jarzynski average were also discussed.

References

1. Sung J (2005) arXiv:cond-mat/0506214
2. Gross DHE (2005) arXiv:cond-mat/0508721
3. Jarzynski C (2005) arXiv:cond-mat/0509344
4. Murashita Y, Funo K, Ueda M (2014) Phys Rev E 90:042110
5. Murashita Y (2015) Absolute irreversibility in information thermodynamics. Master's thesis, The University of Tokyo
6. Jarzynski C (1997) Phys Rev Lett 78:2690
7. Lua RC, Grosberg AY (2005) J Phys Chem B 109:6805
8. Halmos PR (1974) Measure theory. Springer
9. Bartle RG (1995) The elements of integration and lebesgue measure. Wiley
10. Esposito M, Van den Broeck C (2010) Phys Rev Lett 104:090601
11. Seifert U (2012) Rep Prog Phys 75:126001
12. Landauer R (1961) IBM J Res Dev 5:193
13. Hoang DT, Venkatesh BP, Han S, Jo J, Watanabe G, Choi MS (2016) Sci Rep 6:27603

Chapter 4
Fluctuation Theorems in Overdamped Theory with Multiple Reservoirs

As we see in Sect. 2.3, we should be careful when we consider thermodynamics under the overdamped approximation in systems coupled to a thermal environment with position-dependent temperature. In this chapter, we consider another type of non-isothermal systems, namely, systems simultaneously coupled to multiple heat reservoirs with different but constant (i.e., position-independent) temperatures, and address the question of whether the fluctuation theorems hold in the overdamped regime in this non-isothermal system.

First of all, we discuss that naive extensions of the overdamped approximation fail in the presence of multiple heat reservoirs. This is because the velocity degrees of freedom, which are assumed to instantaneously relax in the overdamped approximation, conduct infinite amounts of heat. As a consequence, thermodynamic quantities such as the entropy production diverge, and we cannot therefore deal with this kind of situations if we start from the overdamped Langevin equation or the Fokker-Planck equation. Hence, we need to invoke a sophisticated method to derive the overdamped approximation in the presence of multiple reservoirs. To this aim, we first begin with the underdamped description and construct underdamped stochastic thermodynamics. Then, in the presence of a single reservoir, we show that the overdamped theory can systematically be derived from the underdamped theory by the singular expansion of the time-evolution equation of the generating function. Finally, we derive the overdamped approximation in the presence of multiple reservoirs and show that the fluctuation theorems hold in the ordinary forms for appropriate overdamped thermodynamic quantities. This chapter is mainly based on the published article [1].

Y. Murashita, *Fluctuation Theorems under Divergent Entropy Production and their Applications for Fundamental Problems in Statistical Physics*, Springer Theses, https://doi.org/10.1007/978-981-16-8638-2_4

4.1 Breakdown of Naive Overdamped Approximations with Multiple Reservoirs

As naive attempts to construct an overdamped approximation in the presence of multiple heat reservoirs, we here consider two methods. One is based on the Langevin equation and the other is based on the Fokker-Planck equation.

4.1.1 Naive Overdamped Langevin Approach

We here attempt to construct an overdamped approximation from the overdamped Langevin equation. We consider a situation where one degree of freedom is simultaneously coupled to two heat reservoirs with temperatures T^{h} and T^{c} ($T^{\mathrm{h}} > T^{\mathrm{c}}$). A prototypical example is Feynman's ratchet (see, e.g., Refs. [2, 3]). For simplicity, let us consider a situation in which no external force is applied. Then, the overdamped Langevin equation is written as

$$0 = -\gamma^{\mathrm{h}}\dot{x}_t - \gamma^{\mathrm{c}}\dot{x}_t + \zeta_t^{\mathrm{h}} + \zeta_t^{\mathrm{c}}, \tag{4.1}$$

where γ^{μ}'s ($\mu = \mathrm{h}, \mathrm{c}$) are the friction constants. The white Gaussian noises ζ_t^{μ} ($\mu = \mathrm{h}, \mathrm{c}$) satisfy

$$\langle \zeta_t^{\mu} \rangle = 0, \tag{4.2}$$

$$\langle \zeta_t^{\mu} \zeta_s^{\nu} \rangle = 2\delta^{\mu\nu}\gamma^{\mu} k_{\mathrm{B}} T^{\mu} \delta(t-s). \tag{4.3}$$

Due to the work by Sekimoto [4, 5], the heat flowing from the system to each heat reservoir should be defined as the negative of the "work" that the reservoir does on the system via the friction force and the thermal force. Namely, the heats flowing from time $t = 0$ to τ are defined by

$$Q^{\mu} = -\int_0^{\tau} dt\, (-\gamma^{\mu}\dot{x}_t + \zeta_t^{\mu})\dot{x}_t. \tag{4.4}$$

From the Langevin equation (4.1), the average of the heats can formally be calculated as

$$\langle Q^{\mathrm{h}} \rangle = -\langle Q^{\mathrm{c}} \rangle = -\frac{2k_{\mathrm{B}}\gamma^{\mathrm{h}}\gamma^{\mathrm{c}}}{(\gamma^{\mathrm{h}} + \gamma^{\mathrm{c}})^2}(T^{\mathrm{h}} - T^{\mathrm{c}}) \int_0^{\tau} dt\, \delta(0). \tag{4.5}$$

Thus, $\langle Q^{\mathrm{h}} \rangle$ ($\langle Q^{\mathrm{c}} \rangle$) negatively (positively) diverges. In this way, the naive extension of the overdamped Langevin equation results in ill-defined heats and fails to evaluate thermodynamic quantities.

The physical origin of this divergence lies in the assumption of the overdamped approximation that the velocity relaxation time ε is infinitesimal. Unlike the isothermal case, the velocity is not in an equilibrium state but in a nonequilibrium steady state, and therefore transfer heat in the velocity relaxation time scale ε. Hence, the heat conductivity, which is inversely proportional to ε, diverges.

On the basis of this observation, we can roughly estimate the averaged heat. The velocity relaxation time scale is given by $\varepsilon = m/(\gamma^{\mathrm{h}} + \gamma^{\mathrm{c}})$ with m defined as the mass of the particle. Therefore, the δ function on the right-hand side of Eq. (4.5) is roughly estimated to be $\delta(0) \sim \varepsilon^{-1} = (\gamma^{\mathrm{h}} + \gamma^{\mathrm{c}})/m$. As a result, the averaged amounts of heat are estimated as

$$\langle Q^{\mathrm{h}} \rangle = -\langle Q^{\mathrm{c}} \rangle \sim -\frac{2k_{\mathrm{B}}\gamma^{\mathrm{h}}\gamma^{\mathrm{c}}}{m(\gamma^{\mathrm{h}} + \gamma^{\mathrm{c}})}(T^{\mathrm{h}} - T^{\mathrm{c}})\tau. \tag{4.6}$$

This result is consistent with the direct calculation from the underdamped Langevin equation (see Ref. [3]).

We note that the divergence of heat can be circumvented by introducing additional degrees of freedom. For example, we can consider models consisting of two or more interacting Brownian particles where each particle is coupled to at most one heat reservoir [5–8]. Some models of Feynman's ratchet (e.g., [3, 4, 9, 10]) belong to this category. For these models, the divergence of heat does not occur since each velocity equilibrates with its own reservoir and therefore belong to an equilibrium state.

4.1.2 Naive Fokker-Planck Approach

We here seek to construct overdamped stochastic thermodynamics from the Fokker-Planck equation with the additive current in the presence of multiple reservoirs [11]. In this case, the time evolution of the overdamped probability distribution function $P_t^{\mathrm{od}}(x)$ is written as

$$\partial_t P_t^{\mathrm{od}}(x) = -\partial_x J_t^{\mathrm{od}}(x), \tag{4.7}$$

where the current J_t^{od} can be decomposed into contributions of each reservoir. Namely, the current can be written as

$$J_t^{\mathrm{od}}(x) = \sum_{\mu} J_t^{\mathrm{od},\mu}(x), \tag{4.8}$$

where we define

$$J_t^{\mathrm{od},\mu}(x) = \frac{1}{\gamma^{\mu}}(f_t(x) - k_{\mathrm{B}}T^{\mu}\partial_x)P_t^{\mathrm{od}}(x) \tag{4.9}$$

with $f_t(x)$ being the systematic force. We note that an equivalent dynamics can be derived from the continuous limit of the master equation in the presence of multiple reservoirs [12]. As elaborated in Ref. [11], the averaged heat flowing from the system to the μth reservoir should be identified as

$$\langle \dot{Q}_t^\mu \rangle = \int dx\, J_t^{\mathrm{od},\mu}(x) f_t(x). \tag{4.10}$$

Therefore, the heat flows vanish in the absence of the systematic force. This is an unphysical consequence since the velocity degrees of freedom convey nonzero heat between the reservoirs at different temperatures. In Ref. [11], it is implicitly assumed that the velocity degrees of freedom have no contribution to thermodynamic quantities, but there are various situations where this assumption breaks down in the presence of multiple reservoirs.

4.2 Underdamped Stochastic Thermodynamics

In the previous section, we see that two naive approaches to construct an overdamped approximation fail in the presence of multiple reservoirs. Therefore, we here go back to the underdamped Langevin description and construct stochastic thermodynamics.

4.2.1 Underdamped Langevin Dynamics

We consider an N-dimensional underdamped Langevin system coupled to multiple heat reservoirs. The underdamped Langevin equation reads

$$dx_t = v_t dt, \tag{4.11}$$

$$m dv_t = f_t(x_t)dt + \sum_\mu (-\gamma^\mu v_t dt + \sqrt{2\gamma^\mu k_\mathrm{B} T^\mu} dw_t^\mu), \tag{4.12}$$

where γ^μ and T^μ are the friction coefficient and the temperature of the μth reservoir, respectively. The systematic force $f_t(x)$ can be decomposed into the conservative force and the nonconservative one as

$$f_t(x) = -\partial_x V_t(x) + f_t^{\mathrm{nc}}(x). \tag{4.13}$$

The thermal noises are modelled by the Wiener processes dw_t^μ satisfying

$$\langle dw_t^\mu \rangle = 0, \tag{4.14}$$

$$dw_t^\mu (dw_s^\nu)^T = \delta^{\mu\nu} \mathbf{1} dt, \tag{4.15}$$

where $\mathbf{1}$ is the $N \times N$ unit matrix.

Following Refs. [4, 5], we define heat flowing from the system to the μth reservoir by

$$\delta Q_t^\mu := -v_t \circ (-\gamma^\mu v_t dt + \sqrt{2\gamma^\mu k_B T^\mu} dw_t^\mu) \tag{4.16}$$
$$= \gamma^\mu v_t^2 dt - \sqrt{2\gamma^\mu k_B T^\mu} v_t \circ dw_t^\mu, \tag{4.17}$$

where the symbol $\circ$ represents the Stratonovich product (see Appendix 1 for detail) and the inner product for N-dimensional vectors at the same time. The Stratonovich product can be transformed into the Itô product as

$$\delta Q_t^\mu = \gamma^\mu \left(v_t^2 - \frac{N k_B T^\mu}{m} \right) dt - \sqrt{2\gamma^\mu k_B T^\mu} v_t \cdot dw_t^\mu, \tag{4.18}$$

where the symbol $\cdot$ represents the Itô product and the inner product at the same time. Therefore, the averaged heat is given by

$$\delta \langle Q_t^\mu \rangle = \frac{N k_B \gamma^\mu}{m} \left(\frac{m \langle v_t^2 \rangle}{N k_B} - T^\mu \right) dt, \tag{4.19}$$

which indicates that the heat flows are proportional to the difference between the effective temperature of the velocity degrees of freedom and the physical temperature of the μth heat reservoir.

We note that Eqs. (4.11) and (4.12) are dynamically equivalent to the following equations:

$$dx_t = v_t dt, \tag{4.20}$$
$$m dv_t = f_t(x_t) dt - \gamma^{\text{eff}} v_t dt + \sqrt{2\gamma^{\text{eff}} k_B T^{\text{eff}}} dw_t^{\text{eff}}, \tag{4.21}$$

where we define the effective friction coefficient

$$\gamma^{\text{eff}} = \sum_\mu \gamma^\mu \tag{4.22}$$

and the effective temperature

$$T^{\text{eff}} = \frac{\sum_\mu \gamma^\mu T^\mu}{\sum_\mu \gamma^\mu}, \tag{4.23}$$

and dw_t^{eff} is another Wiener process. However, in this description the heat from each reservoir cannot be distinguished since the thermal noises are mixed up.

4.2.2 *First and Second Laws of Thermodynamics*

We here consider the first and second laws of thermodynamics. The energy of this system is given by

$$U_t(x,t) = V_t(x) + \frac{1}{2}mv^2. \tag{4.24}$$

Therefore, its change can be written as

$$dU_t(x_t, v_t) = (\partial_t V_t(x_t))dt + (\partial_x V_t(x_t))dx_t + mv_t \circ dv_t. \tag{4.25}$$

By inserting Eq. (4.12), we obtain

$$\begin{aligned} dU_t(x_t, v_t) &= (\partial_t V_t(x_t))dt + (\partial_x V_t(x_t))dx_t + v_t f_t(x_t)dt - \sum_\mu (\gamma^\mu v_t^2 dt - \sqrt{2\gamma^\mu k_\mathrm{B} T^\mu} v_t \circ dw_t^\mu) \\ &= \delta W_t^\mathrm{c} + \delta W_t^\mathrm{nc} - \sum_\mu \delta Q_t^\mu \\ &= \delta W_t - \sum_\mu \delta Q_t^\mu, \end{aligned} \tag{4.26}$$

where we define the conservative work

$$\delta W_t^\mathrm{c} := (\partial_t V_t(x_t))dt, \tag{4.27}$$

the nonconservative work

$$\delta W_t^\mathrm{nc} := v_t f_t^\mathrm{nc}(x_t)dt, \tag{4.28}$$

and the total work

$$\delta W_t := \delta W_t^\mathrm{c} + \delta W_t^\mathrm{nc} = (\partial_t V_t(x_t))dt + v_t f_t^\mathrm{nc}(x_t)dt. \tag{4.29}$$

Therefore, by integrating from $t = 0$ to τ, we obtain the first law at the trajectory level as

$$\Delta U = W^\mathrm{c} + W^\mathrm{nc} - \sum_\mu Q^\mu = W - \sum_\mu Q^\mu, \tag{4.30}$$

where $\Delta U = U_\tau(x_\tau, v_\tau) - U_0(x_0, v_0)$ and $I = \int_{t=0}^\tau \delta I_t$ $(I = W^\mathrm{c}, W^\mathrm{nc}, W, Q^\mu)$. Hence, we obtain the first law at the ensemble level as

$$\langle \Delta U \rangle = \langle W^\mathrm{c} \rangle + \langle W^\mathrm{nc} \rangle - \sum_\mu \langle Q^\mu \rangle = \langle W \rangle - \sum_\mu \langle Q^\mu \rangle. \tag{4.31}$$

To formulate the second law of thermodynamics, we define the unaveraged Shannon entropy of the system by

$$s_t := -\ln P_t(x_t, v_t), \tag{4.32}$$

where P_t is the joint probability distribution function in phase space at time t. Since the thermodynamic entropy production in the μth reservoir is given by Q^μ / T^μ, the entropy production of the total system is identified as

$$\Delta s^{\mathrm{tot}} := \Delta s + \sum_{\mu} \frac{Q^{\mu}}{T^{\mu}}, \tag{4.33}$$

where $\Delta s = s_{\tau} - s_0$. The total entropy production can be either positive or negative at the trajectory level. Nevertheless, from the fluctuation theorem derived later, we can derive the second law at the ensemble level

$$\langle \Delta s^{\mathrm{tot}} \rangle \geq 0. \tag{4.34}$$

4.2.3 Generating Function

Suppose that we are interested in a set of quantities in the form

$$\delta X_t^i = a_t^{i(0)}(x_t, v_t)dt + \frac{1}{\varepsilon} a_t^{i(-1)}(x_t, v_t)dt + \sum_{\mu} \sqrt{\frac{2\tilde{\gamma}^{\mu}\tilde{T}^{\mu}}{\varepsilon}} b_t^{i\mu}(x_t, v_t) \cdot dw_t^{\mu}, \tag{4.35}$$

where we define the momentum relaxation time, the rescaled friction coefficient and the rescaled temperature as

$$\varepsilon = \frac{m}{\gamma^{\mathrm{eff}}}, \tag{4.36}$$

$$\tilde{\gamma}^{\mu} = \frac{\gamma^{\mu}}{\gamma^{\mathrm{eff}}}, \tag{4.37}$$

$$\tilde{T}^{\mu} = \frac{T^{\mu}}{T^{\mathrm{eff}}}, \tag{4.38}$$

respectively. We note that these rescaled quantities satisfy

$$\sum_{\mu} \tilde{\gamma}^{\mu} = 1, \tag{4.39}$$

$$\sum_{\mu} \tilde{\gamma}^{\mu}\tilde{T}^{\mu} = 1. \tag{4.40}$$

To investigate the statistics of these quantities we define the generating function as

$$G_{X,t}(x, v, \{\Lambda^i\}) := \left\langle \delta(x_t - x)\delta(v_t - v) \exp\left[\sum_i \Lambda^i X_t^i\right]\right\rangle. \tag{4.41}$$

When all Λ^i vanish, the generating function reduces to the probability distribution function as

$$G_{X,t}(x, v, \{0\}) = \langle \delta(x_t - x)\delta(v_t - v)\rangle =: P_t(x, v). \tag{4.42}$$

When we set $X_0^i = 0$ at the initial time, we obtain the initial condition

$$G_{X,0}(x, v, \{\Lambda^i\}) = \langle \delta(x_0 - x)\delta(v_0 - v)\rangle =: P_0(x, v) \tag{4.43}$$

Moreover, we define the integrated generating function

$$\mathscr{G}_{X,t}(\{\Lambda^i\}) := \left\langle \exp\left[\sum_i \Lambda^i X_t^i\right]\right\rangle = \int dxdv\, G_{X,t}(x, v, \{\Lambda^i\}) \tag{4.44}$$

and the cumulant generating function

$$\mathscr{C}_{X,t}(\{\Lambda^i\}) := \ln\left\langle \exp\left[\sum_i \Lambda^i X_t^i\right]\right\rangle = \ln \mathscr{G}_{X,t}(\{\Lambda^i\}). \tag{4.45}$$

The statistics of $\{X_t^i\}$ can be obtained from these functions. For example, the average is given by

$$\langle X_t^i \rangle = \partial_{\Lambda^i} \mathscr{G}_{X,t}|_{\{\Lambda^k=0\}} = \partial_{\Lambda^i} \mathscr{C}_{X,t}|_{\{\Lambda^k=0\}}, \tag{4.46}$$

and the covariance is given by

$$\langle X_t^i X_t^j \rangle - \langle X_t^i \rangle\langle X_t^j \rangle = \partial_{\Lambda^i}\partial_{\Lambda^j} \mathscr{G}_{X,t}|_{\{\Lambda^k=0\}} - \partial_{\Lambda^i} \mathscr{G}_{X,t}|_{\{\Lambda^k=0\}} \partial_{\Lambda^j} \mathscr{G}_{X,t}|_{\{\Lambda^k=0\}} \tag{4.47}$$

$$= \partial_{\Lambda^i}\partial_{\Lambda^j} \mathscr{C}_{X,t}|_{\{\Lambda^k=0\}}. \tag{4.48}$$

As detailed in Appendix 3, we can derive the time-evolution equation of the generating function as

$$\partial_t G_{X,t}(x, v, \{\Lambda^i\}) = \mathscr{L}_{X,t}(\{\Lambda^i\}) G_{X,t}(x, v, \{\Lambda^i\}). \tag{4.49}$$

The time-evolution operator $\mathscr{L}_{X,t}(\{\Lambda^i\})$ is decomposed as

$$\mathscr{L}_{X,t}(\{\Lambda^i\}) = \mathscr{L}_{X,t}^{(0)}(\{\Lambda^i\}) + \frac{1}{\varepsilon}\mathscr{L}_{X,t}^{(-1)}(\{\Lambda^i\}), \tag{4.50}$$

where we define

$$\mathscr{L}^{(0)}_{X,t}(\{\Lambda^i\}) := -v\cdot\partial_x - \frac{1}{m}f_t(x)\cdot\partial_v + \sum_i \Lambda^i a^{i(0)}, \tag{4.51}$$

$$\begin{aligned}\mathscr{L}^{(-1)}_{X,t}(\{\Lambda^i\}) := \frac{k_B T^{\mathrm{eff}}}{m}\partial_v^2 + \partial_v\cdot v + \sum_i \Lambda^i a^{i(-1)} \\ + \sum_{i,j,\mu}\tilde{\gamma}^\mu \tilde{T}^\mu \Lambda^i\Lambda^j (b^{i\mu}\cdot b^{j\mu}) \\ -2\sqrt{\frac{k_B T^{\mathrm{eff}}}{m}}\sum_{i,\mu}\tilde{\gamma}^\mu\tilde{T}^\mu\Lambda^i\partial_v\cdot b^{i\mu}.\end{aligned} \tag{4.52}$$

When we set $\{\Lambda^i = 0\}$, the time-evolution operator reduces to the generator of the Kramers equation as

$$\mathscr{L}_{X,t}(\{0\}) = \mathscr{L}^{\mathrm{K}}_t := -v\cdot\partial_x - \frac{1}{m}f_t(x)\cdot\partial_v + \frac{1}{\varepsilon}\frac{k_B T^{\mathrm{eff}}}{m}\partial_v^2 + \frac{1}{\varepsilon}\partial_v\cdot v. \tag{4.53}$$

When we are interested in the statistics of heat

$$\delta Q^\mu = \gamma^\mu\left(v_t^2 - \frac{Nk_B T^\mu}{m}\right)dt - \sqrt{2\gamma^\mu k_B T^\mu}v_t\cdot dw_t^\mu, \tag{4.54}$$

we should set

$$a_t^{\mu(0)}(x,v) = 0, \tag{4.55}$$

$$a_t^{\mu(-1)}(x,v) = \frac{m\tilde{\gamma}^\mu}{k_B T^{\mathrm{eff}}}v^2 - N\tilde{\gamma}^\mu\tilde{T}^\mu, \tag{4.56}$$

$$b_t^{\mu\nu}(x,v) = -\delta^{\mu\nu}\sqrt{\frac{m}{k_B T^{\mathrm{eff}}}}v. \tag{4.57}$$

As a result, we obtain the time-evolution operator of the heat-generating function as

$$\mathscr{L}_{Q,t}(\{\Lambda^\mu\}) = \mathscr{L}^{(0)}_t + \frac{1}{\varepsilon}\mathscr{L}^{(-1)}_Q(\{\Lambda^\mu\}), \tag{4.58}$$

where we define

$$\mathscr{L}^{(0)}_t = -v\cdot\partial_x - \frac{1}{m}f_t(x)\cdot\partial_v, \tag{4.59}$$

$$\mathscr{L}^{(-1)}_Q(\{\Lambda^\mu\}) = \frac{k_B T^{\mathrm{eff}}}{m}\partial_v^2 + (1+B)\partial_v\cdot v + Bv\cdot\partial_v + \frac{m}{k_B T^{\mathrm{eff}}}Av^2 \tag{4.60}$$

with

$$A(\{\Lambda^\mu\}) = \sum_\mu \Lambda^\mu \tilde{\gamma}^\mu (1 + \Lambda^\mu \tilde{T}^\mu), \tag{4.61}$$

$$B(\{\Lambda^\mu\}) = \sum_\mu \Lambda^\mu \tilde{\gamma}^\mu \tilde{T}^\mu. \tag{4.62}$$

4.2.4 Fluctuation Theorems

We here enumerate the fluctuation theorems in underdamped stochastic thermodynamics. The detailed derivations are given in Appendix 4 since they are somewhat complicated.

For any initial condition and any time interval, the integral fluctuation theorem of the total entropy production holds:

$$\langle e^{-\Delta s^{\mathrm{tot}}} \rangle = 1 - \lambda, \tag{4.63}$$

where λ is the degree of absolute irreversibility. Therefore the averaged entropy production is always positive as

$$\langle \Delta s^{\mathrm{tot}} \rangle \geq -\ln(1 - \lambda) \geq 0. \tag{4.64}$$

Next, we consider the process that starts from a thermal equilibrium state $P_0^{\mathrm{eq}}(x, v)$ with respect to one of the reservoirs with temperature T^0, where we define the equilibrium probability distribution function

$$P_t^{\mathrm{eq}}(x, v) = \exp\left[-\frac{U_t(x, v) - F_t}{k_{\mathrm{B}} T^0}\right], \tag{4.65}$$

and the free energy

$$F_t = -k_{\mathrm{B}} T^0 \ln \int dx dv \, \exp\left[-\frac{U_t(x, v)}{k_{\mathrm{B}} T^0}\right]. \tag{4.66}$$

We define the irreversible entropy production [13], which is a generalization of the dissipated heat in isothermal systems, as

$$\Delta_{\mathrm{i}} s = -\ln P_\tau^{\mathrm{eq}}(x_\tau, v_\tau) + \ln P_0^{\mathrm{eq}}(x_0, v_0) + \sum_\mu \frac{Q^\mu}{k_{\mathrm{B}} T^\mu} \tag{4.67}$$

$$= \frac{W - \Delta F - \sum_\mu \eta^\mu Q^\mu}{k_{\mathrm{B}} T^0}, \tag{4.68}$$

where $\eta^\mu := 1 - T^0/T^\mu$ is the Carnot efficiency between the reference reservoir and the μth reservoir. For this quantity, the detailed fluctuation theorem holds:

$$\frac{\bar{P}(-\Delta_{\mathrm{i}}s)}{P(\Delta_{\mathrm{i}}s)} = e^{-\Delta_{\mathrm{i}}s}, \tag{4.69}$$

where $\bar{P}$ is the probability of the time-reversed process. We hence obtain the fluctuation theorem

$$\langle e^{-\Delta_{\mathrm{i}}s} \rangle = 1 \tag{4.70}$$

and the second-law-like inequality

$$\langle \Delta_{\mathrm{i}}s \rangle \geq 0. \tag{4.71}$$

Finally, the steady-state fluctuation theorem is satisfied due to the symmetry

$$\mathcal{L}^{\ddagger}_{Q,t}(\{-\Lambda^{\mu} - \tilde{\beta}^{\mu}\}) = \mathcal{L}_{Q,t}(\{\Lambda^{\mu}\}) \tag{4.72}$$

as long as the van Zon-Cohen singularity is absent (see Appendix 4 for detail), where we define $\tilde{\beta}^{\mu} = 1/\tilde{T}^{\mu}$ and the symbol ‡ indicates the inversion of velocities and the Hermitian conjugate at the same time.

4.3 Consistency of the Overdamped Limit in the Isothermal Dynamics

We here consider the system coupled to a single reservoir, and derive the overdamped approximation by the singular expansion of the time-evolution equation of the generating function. In this isothermal setup, the result coincides with the overdamped theory constructed from the overdamped Langevin equation.

4.3.1 Overdamped Theory Derived from the Generating Function

We here approximate the time evolution of the generating function by assuming that we are interested in a time scale much larger than the velocity relaxation time scale $\varepsilon = m/\gamma$. As a result, the velocity degrees of freedom are eliminated and the time evolution with respect to positions is obtained. We note that similar methods are used in Refs. [14–16] to derive the overdamped approximation for systems coupled to a heat reservoir with a position-dependent temperature.

To be specific, we consider the time evolution of the heat-generating function

$$\mathcal{L}_{Q,t}(\Lambda) = \mathcal{L}^{(0)}_{t} + \frac{1}{\varepsilon}\mathcal{L}^{(-1)}_{Q}(\Lambda), \tag{4.73}$$

where

$$\mathscr{L}_t^{(0)} = -v \cdot \partial_x - \frac{1}{m} f_t(x) \cdot \partial_v, \tag{4.74}$$

$$\mathscr{L}_Q^{(-1)}(\Lambda) = \frac{k_B T}{m} \partial_v^2 + (1+\Lambda)\partial_v \cdot v + \Lambda v \cdot \partial_v + \frac{m}{k_B T} \Lambda(1+\Lambda) v^2, \tag{4.75}$$

and we derive its overdamped approximation.

4.3.2 Time-Scale Separation and Singular Expansion

We introduce three time scales: fast time scale $\theta = \varepsilon^{-1} t$, intermediate time scale t, and slow time scale $\hat{t} = \varepsilon t$, and deal them as if they were independent variables to conduct the singular expansion. We assume that ε is small and the three time scales are therefore separated. Moreover, we assume that the time dependence of the systematic force is restricted to the intermediate and slow time scales as $f = f_{t,\hat{t}}$. Then, the time evolution of the heat generating function is given by

$$\frac{1}{\varepsilon} \partial_\theta G_{Q,\theta,t,\hat{t}} + \partial_t G_{Q,\theta,t,\hat{t}} + \varepsilon \partial_{\hat{t}} G_{Q,\theta,t,\hat{t}} = \mathscr{L}_{t,\hat{t}}^{(0)} G_{Q,\theta,t,\hat{t}} + \frac{1}{\varepsilon} \mathscr{L}_Q^{(-1)}(\Lambda). \tag{4.76}$$

We assume that the generating function G can be expanded with respect to ε as

$$G_{Q,\theta,t,\hat{t}} = G_{Q,\theta,t,\hat{t}}^{(0)} + \varepsilon G_{Q,\theta,t,\hat{t}}^{(1)} + \varepsilon^2 G_{Q,\theta,t,\hat{t}}^{(2)} + \cdots. \tag{4.77}$$

Therefore, from each order of ε of Eq. (4.76), we obtain a set of equations as

$$-(\partial_\theta - \mathscr{L}_Q^{(-1)}(\Lambda)) G_{Q,\theta,t,\hat{t}}^{(0)} = 0, \tag{4.78}$$

$$-(\partial_\theta - \mathscr{L}_Q^{(-1)}(\Lambda)) G_{Q,\theta,t,\hat{t}}^{(1)} = (\partial_t - \mathscr{L}_{t,\hat{t}}^{(0)}) G_{Q,\theta,t,\hat{t}}^{(0)}, \tag{4.79}$$

$$-(\partial_\theta - \mathscr{L}_Q^{(-1)}(\Lambda)) G_{Q,\theta,t,\hat{t}}^{(2)} = (\partial_t - \mathscr{L}_{t,\hat{t}}^{(0)}) G_{Q,\theta,t,\hat{t}}^{(1)} + \partial_{\hat{t}} G_{Q,\theta,t,\hat{t}}^{(0)}, \tag{4.80}$$

and higher-order equalities. We solve these equations iteratively from a lower order to a higher order.

The equality of $\mathscr{O}(\varepsilon^{-1})$ is solved once we obtain the eigenfunctions of the operator $\mathscr{L}_Q^{(-1)}(\Lambda)$. To this aim, we transform $\mathscr{L}_Q^{(-1)}(\Lambda)$ into a Hermitian operator as

$$\mathscr{L}_Q^{(-1),H}(\Lambda) := e^{\frac{1+2\Lambda}{4} \frac{mv^2}{k_B T}} \mathscr{L}_Q^{(-1)}(\Lambda) e^{-\frac{1+2\Lambda}{4} \frac{mv^2}{k_B T}} \tag{4.81}$$

$$= \frac{k_B T}{m} \partial_v^2 - \frac{m}{4 k_B T} v^2 + \frac{1}{2} N. \tag{4.82}$$

This time-evolution operator is essentially the negative of the Hamiltonian of the quantum harmonic oscillator. Hence, the eigenvalues are

$$-E_{\{n_i\}} = -\sum_{i=1}^{N} n_i \ (n_i = 0, 1, \ldots) \tag{4.83}$$

and the corresponding eigenfunctions are

$$\psi_{\{n_i\}}(v) = e^{-\frac{mv^2}{4k_BT}} \prod_{i=1}^{N} \sqrt{\frac{1}{2^{n_i} n_i!}\sqrt{\frac{m}{2\pi k_B T}}} H_{n_i}\left(\sqrt{\frac{m}{2k_B T}} v_i\right), \tag{4.84}$$

where $H_n(\cdot)$ is the nth-order Hermite polynomial. Therefore, the original non-Hermite operator $\mathscr{L}_Q^{(-1)}(\Lambda)$ has the eigenvalues

$$-E_{\{n_i\}} = -\sum_{i=1}^{N} n_i, \tag{4.85}$$

the right eigenfunctions

$$\phi_{\{n_i\}}(v, \Lambda) = e^{-\frac{1+\Lambda}{2}\frac{mv^2}{k_BT}} \prod_{i=1}^{N} \sqrt{\frac{1}{2^{n_i} n_i!}\sqrt{\frac{m}{2\pi k_B T}}} H_{n_i}\left(\sqrt{\frac{m}{2k_B T}} v_i\right), \tag{4.86}$$

and the left eigenfunctions

$$\bar{\phi}_{\{n_i\}}(v, \Lambda) = e^{\frac{\Lambda}{2}\frac{mv^2}{k_BT}} \prod_{i=1}^{N} \sqrt{\frac{1}{2^{n_i} n_i!}\sqrt{\frac{m}{2\pi k_B T}}} H_{n_i}\left(\sqrt{\frac{m}{2k_B T}} v_i\right). \tag{4.87}$$

The orthogonality of the eigenfunctions is given by

$$\int dv\, \bar{\phi}_{\{m_i\}}(v, \Lambda)\phi_{\{n_i\}}(v, \Lambda) = \prod_{i=1}^{N} \delta_{m_i n_i}. \tag{4.88}$$

Therefore, the solution of Eq. (4.78) is a sum of the functions of the form

$$e^{-E_{\{n_i\}}\theta}\phi_{\{n_i\}}(v, \Lambda). \tag{4.89}$$

When we are interested in the dynamics in the intermediate time scale t, the term with the smallest $E_{\{n_i\}}$, i.e., $E_{\{0\}} = 0$, survives and the other terms exponentially decay in the fast time scale θ. Then, the generating function can be approximated as

$$G^{(0)}_{Q,\theta,t,\hat{t}}(x, v, \Lambda) = \hat{G}^{(0)}_{Q,t,\hat{t}}(x, \Lambda)\phi_{\{0\}}(v, \Lambda), \tag{4.90}$$

where $\hat{G}^{(0)}_{Q,t,\hat{t}}(x, \Lambda)$ is an arbitrary function independent of θ and v. Thus, the dependence on the positional degrees of freedom is separated from the velocity degrees of freedom. In a similar way, as long as we are not interested in the evolution in the fast time scale, we can replace the fast time derivative ∂_θ with the largest eigenvalue of $\mathscr{L}^{(-1)}_Q(\Lambda)$, i.e., with zero. Therefore, the time-evolution equations reduce to

$$\mathscr{L}^{(-1)}_Q(\Lambda)G^{(1)}_{Q,\theta,t,\hat{t}}(x, v, \Lambda) = (\partial_t - \mathscr{L}^{(0)}_{t,\hat{t}})G^{(0)}_{Q,\theta,t,\hat{t}}(x, v, \Lambda), \tag{4.91}$$

$$\mathscr{L}^{(-1)}_Q(\Lambda)G^{(2)}_{Q,\theta,t,\hat{t}}(x, v, \Lambda) = (\partial_t - \mathscr{L}^{(0)}_{t,\hat{t}})G^{(1)}_{Q,\theta,t,\hat{t}}(x, v, \Lambda) + \partial_{\hat{t}}G^{(0)}_{Q,\theta,t,\hat{t}}(x, v, \Lambda). \tag{4.92}$$

4.3.3 Consistency Relations

By multiplying Eqs. (4.79) and (4.80) by $\bar{\phi}_{\{0\}}$ and integrating them over v, we obtain

$$0 = \int dv\, \bar{\phi}_{\{0\}}(v, \Lambda)(\partial_t - \mathscr{L}^{(0)}_{t,\hat{t}})G^{(0)}_{Q,\theta,t,\hat{t}}(x, v, \Lambda), \tag{4.93}$$

$$0 = \int dv\, \bar{\phi}_{\{0\}}(v, \Lambda)\left[(\partial_t - \mathscr{L}^{(0)}_{t,\hat{t}})G^{(1)}_{Q,\theta,t,\hat{t}}(x, v, \Lambda) + \partial_{\hat{t}}G^{(0)}_{Q,\theta,t,\hat{t}}(x, v, \Lambda)\right], \tag{4.94}$$

since $\bar{\phi}_{\{0\}}$ is the left eigenfunction of $\mathscr{L}^{(-1)}_Q(\Lambda)$ with the zero eigenvalue. We call these equations the consistency relations since they should be satisfied for the singular expansion to be consistent. By inserting Eq. (4.90) into Eq. (4.93), we obtain

$$\begin{aligned}
0 &= \int dv\, \bar{\phi}_{\{0\}}(v, \Lambda)\left[\partial_t + v\cdot\left(\partial_x - \frac{1+\Lambda}{k_{\mathrm{B}}T}f_{t,\hat{t}}(x)\right)\right]\hat{G}^{(0)}_{Q,t,\hat{t}}(x, \Lambda)\phi_{\{0\}}(v, \Lambda) \\
&= \partial_t\hat{G}^{(0)}_{Q,t,\hat{t}}(x, \Lambda)\int dv\, \bar{\phi}_{\{0\}}(v, \Lambda)\phi_{\{0\}}(v, \Lambda) \\
&\quad + \left(\partial_x - \frac{1+\Lambda}{k_{\mathrm{B}}T}f_{t,\hat{t}}(x)\right)\hat{G}^{(0)}_{Q,t,\hat{t}}(x, \Lambda)\cdot\int dv\, \bar{\phi}_{\{0\}}(v, \Lambda)v\phi_{\{0\}}(v, \Lambda) \\
&= \partial_t\hat{G}^{(0)}_{Q,t,\hat{t}}(x, \Lambda).
\end{aligned} \tag{4.95}$$

Therefore, Eq. (4.79) reduces to

$$\mathscr{L}^{(-1)}_Q(\Lambda)G^{(1)}_{Q,\theta,t,\hat{t}}(x, v, \Lambda) = v\cdot\left[\left(\partial_x - \frac{1+\Lambda}{k_{\mathrm{B}}T}f_{t,\hat{t}}(x)\right)\hat{G}^{(0)}_{Q,t,\hat{t}}(x, \Lambda)\right]\phi_{\{0\}}(v, \Lambda). \tag{4.96}$$

By noting $\mathscr{L}^{(-1)}_Q(\Lambda)(v_i\phi_{\{0\}}) = -v_i\phi_{\{0\}}$, we obtain the solution

$$G^{(1)}_{Q,\theta,t,\hat{t}}(x, v, \Lambda) = \left[\hat{G}^{(1)}_{Q,t,\hat{t}}(x, \Lambda) - v\cdot\left(\partial_x - \frac{1+\Lambda}{k_{\mathrm{B}}T}f_{t,\hat{t}}(x)\right)\hat{G}^{(0)}_{Q,t,\hat{t}}(x, \Lambda)\right]\phi_{\{0\}}(v, \Lambda), \tag{4.97}$$

where $\hat{G}^{(1)}_{Q,t,\hat{t}}(x,\Lambda)$ is an arbitrary function independent of θ and v. Then, after some algebra, the consistency relation (4.94) reduces to

$$\partial_{\hat{t}}\hat{G}^{(0)}_{Q,t,\hat{t}}(x,\Lambda)+\partial_t\hat{G}^{(1)}_{Q,t,\hat{t}}(x,\Lambda)$$
$$=\left[\frac{k_BT}{m}\partial_x^2-\frac{1+\Lambda}{m}\partial_x\cdot f_{t,\hat{t}}(x)-\frac{\Lambda}{m}f_{t,\hat{t}}(x)\cdot\partial_x+\frac{\Lambda(1+\Lambda)}{mk_BT}f_{t,\hat{t}}(x)^2\right]\hat{G}^{(0)}_{Q,t,\hat{t}}(x,\Lambda). \quad (4.98)$$

4.3.4 Overdamped Approximation

Let us define the overdamped generating function by

$$G^{\mathrm{od}}_{Q,t}(x,\Lambda)=\left(\sqrt{\frac{1}{1+\Lambda}}\sqrt{\frac{2\pi k_BT}{m}}\right)^N[\hat{G}^{(0)}_{Q,t,\varepsilon t}(x,\Lambda)+\varepsilon\hat{G}^{(1)}_{Q,t,\varepsilon t}(x,\Lambda)]. \quad (4.99)$$

Then, from Eq. (4.98), the time-evolution of this function is given by

$$\partial_t G^{\mathrm{od}}_{Q,t}(x,\Lambda)\simeq\mathscr{L}^{\mathrm{od}}_{Q,t}(\Lambda)G^{\mathrm{od}}_{Q,t}(x,\Lambda), \quad (4.100)$$

where we define

$$\mathscr{L}^{\mathrm{od}}_{Q,t}(\Lambda)=\varepsilon\left[\frac{k_BT}{m}\partial_x^2-\frac{1+\Lambda}{m}\partial_x\cdot f_t(x)-\frac{\Lambda}{m}f_t(x)\cdot\partial_x+\frac{\Lambda(1+\Lambda)}{mk_BT}f_t(x)^2\right] \quad (4.101)$$

and the symbol $\simeq$ represents the equality up to $\mathscr{O}(\varepsilon^2)$. Moreover, by integrating the underdamped heat-generating function over v, we obtain

$$\int dv\, G_{Q,t}(x,v,\Lambda)\simeq G^{\mathrm{od}}_{Q,t}(x,\Lambda). \quad (4.102)$$

Therefore, when Λ vanishes, we obtain

$$G^{\mathrm{od}}_{Q,t}(x,0)\simeq\int dv\, G_{Q,t}(x,v,0)=\int dv\, P_t(x,v)=P^{\mathrm{od}}_t(x). \quad (4.103)$$

Furthermore, the initial condition is given by

$$G^{\mathrm{od}}_{Q,0}(x,\Lambda)\simeq\int dv\, G_{Q,0}(x,v,\Lambda)=\int dv\, P_0(x,v)=:P^{\mathrm{od}}_0(x). \quad (4.104)$$

Thus, from Eqs. (4.100) and (4.104), we can approximately calculate the time evolution of the overdamped generating function (4.99) only by the positional degrees of freedom. By integrating Eq. (4.102) over x, we obtain

$$\mathscr{G}_{Q,t}(\Lambda) \simeq \mathscr{G}^{\rm od}_{Q,t}(\Lambda), \tag{4.105}$$

where we define

$$\mathscr{G}^{\rm od}_{Q,t}(\Lambda) = \int dx\, G^{\rm od}_{Q,t}(x, \Lambda). \tag{4.106}$$

In this way, the overdamped theory can be obtained by way of the time-scale separation and the singular expansion.

4.3.5 Overdamped Theory Derived from the Langevin Equation

We here start from the overdamped Langevin equation and derive the overdamped theory in the presence of a single reservoir. The result coincides with the one derived from the generating function in the previous section.

We consider the N-dimensional overdamped Langevin equation:

$$0 = f_t(x_t)dt - \gamma dx_t + \sqrt{2\gamma k_{\rm B}T}dw_t, \tag{4.107}$$

or equivalently,

$$dx_t = \frac{f_t(x_t)}{\gamma}dt + \sqrt{\frac{2k_{\rm B}T}{\gamma}}dw_t. \tag{4.108}$$

The stochastic heat is defined by [4, 5]

$$\delta Q^{\rm od}_t := -\left(-\gamma\frac{dx_t}{dt} + \sqrt{2\gamma k_{\rm B}T}\frac{dw_t}{dt}\right) \circ dx_t \tag{4.109}$$

$$= f_t(x_t) \circ dx_t. \tag{4.110}$$

Let us consider a quantity of the form

$$\delta X_t = a^{(0)}_t(x_t)dt + \frac{\varepsilon}{mk_{\rm B}T}a^{(1)}_t(x_t)dt + \sqrt{\frac{2\varepsilon}{mk_{\rm B}T}}b_t(x_t)\cdot dw_t. \tag{4.111}$$

The generating function is defined by

$$G^{\rm od}_{X,t}(x, \Lambda) := \langle \delta(x_t - x)\exp[\Lambda X_t]\rangle. \tag{4.112}$$

When $\Lambda = 0$, the generating function reduces to the probability distribution function of the position as

$$G^{\rm od}_{X,t}(x, 0) = \langle \delta(x_t - x)\rangle =: P^{\rm od}_t(x). \tag{4.113}$$

When X_t vanishes at the initial time $t = 0$, we obtain the initial condition

$$G^{\rm od}_{X,0}(x, \Lambda) = \langle \delta(x_0 - x) \rangle = P^{\rm od}_0(x). \tag{4.114}$$

As in the underdamped case, we can derive the time-evolution equation for the generating function as

$$\partial_t G^{\rm od}_{X,t}(x, \Lambda) = \mathscr{L}^{\rm od}_{X,t}(\Lambda) G^{\rm od}_{X,t}(x, \Lambda), \tag{4.115}$$

where we define the time-evolution operator by

$$\begin{aligned} \mathscr{L}^{\rm od}_{X,t}(\Lambda) &= \Lambda a^{(0)}_t(x) \\ &+ \varepsilon \left[\frac{k_{\rm B}T}{m} \partial_x^2 - \frac{1}{m} \partial_x \cdot [f_t(x) + 2\Lambda b_t(x)] + \frac{1}{m k_{\rm B} T} \Lambda [a^{(1)}_t(x) + \Lambda b_t(x)^2] \right]. \end{aligned} \tag{4.116}$$

When we set $\Lambda = 0$, this time-evolution operator reduces to the generator of the Fokker-Planck equation as

$$\mathscr{L}^{\rm od}_{X,t}(0) = \mathscr{L}^{\rm FP}_t := \frac{k_{\rm B}T}{m} \partial_x^2 - \frac{1}{m} \partial_x \cdot f_t(x). \tag{4.117}$$

Since the heat is given by

$$\delta Q^{\rm od}_t = \frac{1}{\gamma k_{\rm B} T} \left[f_t(x_t)^2 + k_{\rm B} T \partial_x (f_t(x_t)) \right] dt + \sqrt{\frac{2}{\gamma k_{\rm B} T}} f_t(x_t) \cdot dw_t, \tag{4.118}$$

the time evolution of the heat-generating function is determined as

$$\begin{aligned} \mathscr{L}^{\rm od}_{Q,t}(\Lambda) &= \frac{k_{\rm B}T}{m} \partial_x^2 - \frac{1}{m} \partial_x \cdot [f_t(x) + 2\Lambda f_t(x)] \\ &+ \frac{1}{m k_{\rm B} T} \Lambda [f_t(x)^2 + k_{\rm B} T \partial_x (f_t(x)) + \Lambda f_t(x)^2] \\ &= \frac{k_{\rm B}T}{m} \partial_x^2 - \frac{1+\Lambda}{m} \partial_x \cdot f_t(x) - \frac{\Lambda}{m} f_t(x) \cdot \partial_x + \frac{\Lambda(1+\Lambda)}{m k_{\rm B} T} f_t(x)^2, \end{aligned} \tag{4.119}$$

which coincides with the time-evolution operator (4.101) derived in the previous section. Therefore, the overdamped theory constructed from the overdamped Langevin equation is equivalent to that derived from the heat-generating function. In a similar way, we can show the equivalence of the two methods for the total entropy production $\Delta s^{\rm tot}$ and the irreversible entropy production $\Delta_{\rm i} s$ as shown in Appendix 5.

4.4 Overdamped Theory with Multiple Reservoirs

We here consider the overdamped approximation in the presence of multiple heat reservoirs. As we see in Sect. 4.1, we cannot construct the overdamped theory when we naively start from the overdamped Langevin equation. Therefore, we start from the underdamped stochastic thermodynamics in Sect. 4.2 and derive its overdamped approximation by using the singular expansion described in Sect. 4.3.

To be specific, we consider the heat-generating function $G_{Q,t}(x, v, \{\Lambda^\mu\})$ and its time evolution operator

$$\mathscr{L}_{Q,t}(\{\Lambda^\mu\}) = \mathscr{L}_t^{(0)} + \frac{1}{\varepsilon}\mathscr{L}_Q^{(-1)}(\{\Lambda^\mu\}), \tag{4.120}$$

where

$$\mathscr{L}_t^{(0)} = -v \cdot \partial_x - \frac{1}{m} f_t(x) \cdot \partial_v, \tag{4.121}$$

$$\mathscr{L}_Q^{(-1)}(\{\Lambda^\mu\}) = \frac{k_B T^{\mathrm{eff}}}{m}\partial_v^2 + (1+B)\partial_v \cdot v + Bv \cdot \partial_v + \frac{m}{k_B T^{\mathrm{eff}}} A v^2 \tag{4.122}$$

with

$$A(\{\Lambda^\mu\}) = \sum_\mu \Lambda^\mu \tilde{\gamma}^\mu (1 + \Lambda^\mu \tilde{T}^\mu), \tag{4.123}$$

$$B(\{\Lambda^\mu\}) = \sum_\mu \Lambda^\mu \tilde{\gamma}^\mu \tilde{T}^\mu. \tag{4.124}$$

4.4.1 Overdamped Approximation with Multiple Reservoirs

By assuming the time-scale separation, we obtain the same set of Eqs. (4.78), (4.79) and (4.80). The crucial difference from the previous case with a single reservoir is that the largest eigenvalue of $\mathscr{L}_Q^{(-1)}(\{\Lambda^\mu\})$ is nonzero. Actually, $\mathscr{L}_Q^{(-1)}(\{\Lambda^\mu\})$ can be Hermitianized as

$$\mathscr{L}_Q^{(-1),H}(\{\Lambda^\mu\}) := e^{\frac{1+2B}{4}\frac{mv^2}{k_B T^{\mathrm{eff}}}} \mathscr{L}_Q^{(-1)}(\Lambda) e^{-\frac{1+2B}{4}\frac{mv^2}{k_B T^{\mathrm{eff}}}} \tag{4.125}$$

$$= \frac{k_B T^{\mathrm{eff}}}{m}\partial_v^2 - \frac{m}{k_B T^{\mathrm{eff}}}\frac{(1+2B)^2 - 4A}{4} v^2 + \frac{1}{2}N. \tag{4.126}$$

Therefore, the eigenvalues are

$$-E_{\{n_i\}}(\{\Lambda^\mu\}) = -R\sum_i n_i + \frac{1}{2}N(1-R), \tag{4.127}$$

where we define

$$R(\{\Lambda^\mu\}) := \sqrt{(1+2B)^2 - 4A}. \tag{4.128}$$

The corresponding eigenfunctions are given by

$$\psi_{\{n_i\}}(v, \{\Lambda^\mu\}) = e^{-\frac{R}{4}\frac{mv^2}{k_B T^{\mathrm{eff}}}} \prod_{i=1}^{N} \sqrt{\frac{1}{2^{n_i} n_i!}\sqrt{\frac{mR}{2\pi k_B T^{\mathrm{eff}}}}} H_{n_i}\left(\sqrt{\frac{mR}{2k_B T^{\mathrm{eff}}}} v_i\right). \tag{4.129}$$

Hence, we obtain the largest eigenvalue of $\mathscr{L}_Q^{(-1)}(\{\Lambda^\mu\})$

$$\alpha(\{\Lambda^\mu\}) := \frac{1}{2}N(1-R), \tag{4.130}$$

the corresponding right eigenfunction

$$\phi_\alpha(v, \{\Lambda^\mu\}) = \left(\frac{mR}{2\pi k_B T^{\mathrm{eff}}}\right)^{\frac{N}{4}} e^{-\frac{\kappa}{2}\frac{mv^2}{k_B T^{\mathrm{eff}}}}, \tag{4.131}$$

and the corresponding left eigenfunction

$$\bar{\phi}_\alpha(v, \{\Lambda^\mu\}) = \left(\frac{mR}{2\pi k_B T^{\mathrm{eff}}}\right)^{\frac{N}{4}} e^{-\frac{\rho}{2}\frac{mv^2}{k_B T^{\mathrm{eff}}}}, \tag{4.132}$$

where we define

$$\kappa(\{\Lambda^\mu\}) := \frac{1+2B+R}{2}, \tag{4.133}$$

$$\rho(\{\Lambda^\mu\}) := \frac{-1-2B+R}{2}. \tag{4.134}$$

Since we are interested in the time scale slower than θ, the derivative ∂_θ can be replaced with the largest eigenvalue α. Moreover, because α is nonzero, the generating function has an explicit θ-dependence as

$$G^{(0)}_{Q,\theta,t,\hat{t}}(x, v, \{\Lambda^\mu\}) = \hat{G}^{(0)}_{Q,t,\hat{t}}(x, \{\Lambda^\mu\}) e^{\alpha\theta} \phi_\alpha(v, \{\Lambda^\mu\}). \tag{4.135}$$

This new factor $e^{\alpha\theta}$ corresponds to heat transfer due to the velocity degrees of freedom in the fast time scale. From the consistency relation of $\mathscr{O}(\varepsilon^0)$, we obtain

$$\partial_t \hat{G}^{(0)}_{Q,t,\hat{t}}(x, \{\Lambda^\mu\}) = 0. \tag{4.136}$$

Then, the equation to the order of $\mathscr{O}(\varepsilon^0)$ is solved to be

$$G^{(1)}_{Q,\theta,t,\hat{t}}(x,v,\{\Lambda^{\mu}\})$$
$$= \left[\hat{G}^{(1)}_{Q,t,\hat{t}}(x,\{\Lambda^{\mu}\}) - \frac{1}{R}v\cdot\left(\partial_x - \frac{\kappa}{k_{\mathrm{B}}T^{\mathrm{eff}}} f_{t,\hat{t}}(x)\right)\hat{G}^{(0)}_{Q,t,\hat{t}}(x,\{\Lambda^{\mu}\})\right] e^{\alpha\theta}\phi_{\alpha}(v,\{\Lambda^{\mu}\}). \tag{4.137}$$

Inserting this result into the consistency relation of $\mathscr{O}(\varepsilon)$, we obtain

$$\partial_{\hat{t}}\hat{G}^{(0)}_{Q,t,\hat{t}}(x,\{\Lambda^{\mu}\}) + \partial_t\hat{G}^{(1)}_{Q,t,\hat{t}}(x,\{\Lambda^{\mu}\})$$
$$= \frac{1}{R^2}\left[\frac{k_{\mathrm{B}}T^{\mathrm{eff}}}{m}\partial_x^2 - \frac{\kappa}{m}\partial_x\cdot f_{t,\hat{t}}(x) + \frac{\rho}{m}f_{t,\hat{t}}(x)\cdot\partial_x + \frac{A}{mk_{\mathrm{B}}T^{\mathrm{eff}}}f_{t,\hat{t}}(x)^2\right]\hat{G}^{(0)}_{Q,t,\hat{t}}(x,\{\Lambda^{\mu}\}). \tag{4.138}$$

We define the overdamped generating function

$$G^{\mathrm{od}}_{Q,t}(x,\{\Lambda\}) := \left(\sqrt{\frac{\sqrt{R}}{\kappa}\sqrt{\frac{2\pi k_{\mathrm{B}}T^{\mathrm{eff}}}{m}}}\right)^{N}[\hat{G}^{(0)}_{Q,t,\varepsilon t}(x,,\{\Lambda\}) + \varepsilon\hat{G}^{(1)}_{Q,t,\varepsilon t}(x,,\{\Lambda\})]. \tag{4.139}$$

Then, its time evolution is given by

$$\partial_t G^{\mathrm{od}}_{Q,t}(x,\{\Lambda\}) \simeq \mathscr{L}^{\mathrm{od}}_{Q,t}(\{\Lambda^{\mu}\})G^{\mathrm{od}}_{Q,t}(x,\{\Lambda\}), \tag{4.140}$$

where we define the overdamped time-evolution operator by

$$\mathscr{L}^{\mathrm{od}}_{Q,t}(\{\Lambda^{\mu}\}) := \frac{\varepsilon}{R^2}\left[\frac{k_{\mathrm{B}}T^{\mathrm{eff}}}{m}\partial_x^2 - \frac{\kappa}{m}\partial_x\cdot f_t(x) + \frac{\rho}{m}f_t(x)\cdot\partial_x + \frac{A}{mk_{\mathrm{B}}T^{\mathrm{eff}}}f_t(x)^2\right]. \tag{4.141}$$

By integrating out the velocities in the underdamped generating function, we obtain

$$\int dv\, G_{Q,t}(x,v,\{\Lambda^{\mu}\}) = [G^{\mathrm{od}}_{Q,t}(x,\{\Lambda^{\mu}\}) + \mathscr{O}(\varepsilon^2)]\mathscr{G}^{v}_{Q,t}(\{\Lambda^{\mu}\}), \tag{4.142}$$

where we define the contribution of the velocity degrees of freedom by

$$\mathscr{G}^{v}_{Q,t}(\{\Lambda^{\mu}\}) := \exp\left[\frac{\alpha t}{\varepsilon}\right] = \exp\left[\frac{N(1-R)t}{2\varepsilon}\right]. \tag{4.143}$$

By setting all Λ^{μ} to zero, we obtain

$$G^{\mathrm{od}}_{Q,t}(x,\{0\}) \simeq \int dv\, G_{Q,t}(x,v,\{0\}) = \int dv\, P_t(x,v) = P^{\mathrm{od}}_t(x). \tag{4.144}$$

The initial condition is given by

$$G^{\mathrm{od}}_{Q,0}(x,\{\Lambda^{\mu}\}) \simeq \int dv\, G_{Q,0}(x,v,\{\Lambda^{\mu}\}) = \int dv\, P_0(x,v) = P^{\mathrm{od}}_0(x). \tag{4.145}$$

By integrating Eq. (4.142) over x, we obtain

$$\mathscr{G}_{Q,t}(\{\Lambda^\mu\}) = [\mathscr{G}^{\rm od}_{Q,t}(\{\Lambda^\mu\}) + \mathscr{O}(\varepsilon^2)]\mathscr{G}^{v}_{Q,t}(\{\Lambda^\mu\}), \tag{4.146}$$

where we define the overdamped generating function by

$$\mathscr{G}^{\rm od}_{Q,t}(\{\Lambda^\mu\}) := \int dx\, G^{\rm od}_{Q,t}(x, \{\Lambda^\mu\}). \tag{4.147}$$

We note that similar overdamped approximations can be derived for the total entropy production $\Delta s^{\rm tot}$ and the irreversible entropy production $\Delta s_{\rm i}$ as shown in Appendix 5.

Let us consider the errors of the overdamped approximation. The nth moment of the heat is given by the nth derivative of Eq. (4.146). The largest error originates from the term of the form

$$\mathscr{O}(\varepsilon^2)\frac{\partial^n}{\partial\Lambda^{\mu_1}\cdots\partial\Lambda^{\mu_n}}\exp\left[\frac{\alpha t}{\varepsilon}\right]. \tag{4.148}$$

Thus, the nth moment calculated from the overdamped approximation has an error of $\mathscr{O}(\varepsilon^{2-n})$. The cumulant generating function is approximated as

$$\mathscr{C}_{Q,t}(\{\Lambda^\mu\}) \simeq \frac{\alpha(\{\Lambda^\mu\})}{\varepsilon}t + \ln\mathscr{G}^{\rm od}_{Q,t}(\{\Lambda^\mu\}). \tag{4.149}$$

The first term represents the heat transfer by the velocity degrees of freedom in the fast time scale, while the second term originates from the heat transfer by the positional degrees of freedom in the slower time scales.

We now discuss how to implement the boundary conditions. Let us consider the reflecting boundary condition. In the underdamped theory, the boundary condition at the position x_0 is given by

$$G_{Q,t}(x_0, v, \{\Lambda^\mu\}) = G_{Q,t}(x_0, \mathscr{R}v, \{\Lambda^\mu\}), \tag{4.150}$$

where $\mathscr{R}v := v - 2(v\cdot\hat{n})\hat{n}$ is the velocity reflected by the boundary with the outward-facing normal unit vector $\hat{n}$. The zeroth-order term (4.135) automatically satisfies the boundary condition. For the first-order term (4.137) to satisfy this condition, the following relation should hold

$$\hat{n}\cdot\left(\partial_x - \frac{\kappa}{k_{\rm B}T^{\rm eff}}f_{t,\hat{t}}(x)\right)G^{\rm od}_{Q,t}(x_0, \{\Lambda^\mu\}) = 0, \tag{4.151}$$

which is the approximated reflecting boundary condition. Meanwhile, the absorbing boundary condition $G_{Q,t}(x_0, v, \{\Lambda^\mu\}) = 0$ is approximated as $G^{\rm od}_{Q,t}(x_0, \{\Lambda^\mu\}) = 0$.

Finally, let us compare our method with the two naive approaches in Sec. 4.1. Starting from the overdamped Langevin equation means that we set ε to zero from the beginning. As a result, the heat transferred by the velocity degrees of freedom diverges. In contrast, in our method, we keep ε small but nonzero. This is the reason why we can circumvent the divergence of heat and construct the overdamped theory.

In the naive approach based on the additive Fokker-Planck equation, it is assumed that the fast degrees of freedom do not contribute to the evolution of thermodynamic quantities. This assumption obviously breaks down in our setup since the velocity degrees of freedom transfer heat. Nevertheless, one may think that overdamped theory can be constructed by adding this contribution by hand. Unfortunately, this seems impossible since the time-evolution operator (4.141) cannot be written as the sum of a contribution from each reservoir. This nonadditive form indicates that the heat transfer by one reservoir is nontrivially correlated to that by another reservoir. We thus believe that our systematic procedure is indispensable to derive the overdamped theory.

4.4.2 First and Second Laws of Thermodynamics

We here consider the first and second laws in the overdamped approximation. We define the overdamped stochastic heats $Q_t^{\mathrm{od},\mu}$ as quantities satisfying

$$G_{Q,t}^{\mathrm{od}}(x,\{\Lambda^\mu\}) = \left\langle \delta(x_t - x)\exp\left[\sum_\mu \frac{\Lambda^\mu Q_t^{\mu,\mathrm{od}}}{k_{\mathrm{B}}T^{\mathrm{eff}}}\right]\right\rangle. \tag{4.152}$$

We also define the stochastic heat $Q_t^{v,\mu}$ transferred by the velocity degrees of freedom as

$$\mathscr{G}_{Q,t}^{v}(\{\Lambda^\mu\}) = \left\langle \exp\left[\sum_\mu \frac{\Lambda^\mu Q_t^{\mu,v}}{k_{\mathrm{B}}T^{\mathrm{eff}}}\right]\right\rangle. \tag{4.153}$$

By noting $\alpha(\{\Lambda^\mu = \Lambda\}) = 0$, we obtain

$$\mathscr{G}_{Q,t}^{v}(\{\Lambda^\mu = \Lambda\}) = \left\langle \exp\left[\frac{\Lambda}{k_{\mathrm{B}}T^{\mathrm{eff}}}\sum_\mu Q_t^{\mu,v}\right]\right\rangle = 1 \tag{4.154}$$

for an arbitrary Λ. Hence, we obtain

$$\sum_\mu Q_t^{\mu,v} = 0. \tag{4.155}$$

Thus, the heats transferred by the velocity degrees of freedom in the fast time scale completely balance each other without any contributions of work and heat transfer in slower time scales. At the same time, from Eq. (4.146), we obtain

$$\mathcal{G}_t(\{\Lambda^\mu = \Lambda\}) \simeq \mathcal{G}_t^{\mathrm{od}}(\{\Lambda^\mu = \Lambda\}). \tag{4.156}$$

Therefore, the statics of the sum of the underdamped heat are approximately identical to that of the overdamped heat as

$$\sum_\mu Q_t^\mu \simeq \sum_\mu Q_t^{\mu,\mathrm{od}}, \tag{4.157}$$

which indicates that the net heat current consists only of the overdamped contributions. Consequently, the first law of thermodynamics in the overdamped approximation reads

$$\Delta U \simeq W - \sum_\mu Q_t^{\mu,\mathrm{od}}. \tag{4.158}$$

To formulate the second law of thermodynamics, we define the unaveraged overdamped Shannon entropy by

$$s_t^{\mathrm{od}} := -\ln P_t^{\mathrm{od}}(x_t). \tag{4.159}$$

Then, the total entropy production is identified to be

$$\Delta s^{\mathrm{tot,od}} := \Delta s^{\mathrm{od}} + \sum_\mu \frac{Q^{\mu,\mathrm{od}}}{k_\mathrm{B} T^\mu}, \tag{4.160}$$

where $\Delta s^{\mathrm{od}} := s_\tau^{\mathrm{od}} - s_0^{\mathrm{od}}$. From the fluctuation theorem below, we can show that the second law is satisfied as

$$\langle \Delta s^{\mathrm{tot,od}} \rangle \geq 0. \tag{4.161}$$

4.4.3 Fluctuation Theorems

We here show that the fluctuation theorems in the original underdamped theory remain valid under the overdamped approximation. The derivations are given in Appendix 6.

For any initial condition, we can derive the integral fluctuation theorem for the total entropy production as

$$\langle e^{-\Delta s^{\mathrm{tot,od}}} \rangle = 1 - \lambda, \tag{4.162}$$

where λ is the degree of absolute irreversibility. This is the overdamped analog of Eq. (4.63). Consequently, the second law of thermodynamics is satisfied:

$$\langle \Delta s^{\mathrm{tot,od}} \rangle \geq -\ln(1-\lambda) \geq 0. \tag{4.163}$$

Next, we consider the process that starts from a thermal equilibrium state $P_0^{\mathrm{eq,od}}(x)$ with respect to one of the reservoirs at temperature T^0. We here define

$$P_t^{\mathrm{eq,od}}(x) = \exp\left[-\frac{V_t(x) - F_t^{\mathrm{od}}}{k_\mathrm{B} T^0}\right], \tag{4.164}$$

and

$$F_t^{\mathrm{od}} = -k_\mathrm{B} T^0 \ln \int dx \, \exp\left[-\frac{V_t(x)}{k_\mathrm{B} T^0}\right]. \tag{4.165}$$

In this case, the irreversible entropy production is defined by

$$\Delta_\mathrm{i} s^{\mathrm{od}} = \frac{W - \Delta F^{\mathrm{od}} - \sum_\mu \eta^\mu Q^{\mu,\mathrm{od}}}{k_\mathrm{B} T^0} \tag{4.166}$$

with $\Delta F^{\mathrm{od}} = F_\tau^{\mathrm{od}} - F_0^{\mathrm{od}}$. Then, the detailed fluctuation theorem is satisfied as

$$\frac{\bar{P}(-\Delta_\mathrm{i} s^{\mathrm{od}})}{P(\Delta_\mathrm{i} s^{\mathrm{od}})} = e^{-\Delta_\mathrm{i} s^{\mathrm{od}}}, \tag{4.167}$$

which is the overdamped version of Eq. (4.69). Therefore, we obtain the integral fluctuation theorem

$$\langle e^{-\Delta_\mathrm{i} s^{\mathrm{od}}} \rangle = 1, \tag{4.168}$$

and the second-law-like inequality

$$\langle \Delta_\mathrm{i} s^{\mathrm{od}} \rangle \geq 0. \tag{4.169}$$

Finally, the time-evolution operator has the symmetry of the steady-state fluctuation theorem as

$$\mathcal{L}_{Q,t}^{\mathrm{od},\dagger}(\{-\Lambda^\mu - \tilde{\beta}^\mu\}) = \mathcal{L}_{Q,t}^{\mathrm{od}}(\{\Lambda^\mu\}), \tag{4.170}$$

which is the counterpart of Eq. (4.72).

Thus, the fluctuation theorems remain valid under the overdamped approximation in the presence of the multiple reservoirs after we subtract the contributions to the heat and the entropy productions from the velocity degrees of freedom of $\mathcal{O}(\varepsilon^{-1})$.

Appendix

A.1 Stochastic Integrals

Here, we briefly review stochastic integrals. This section is partly based on Ref. [5]. The Wiener process w_t is a mathematical model of the white Gaussian noise. We define an infinitesimal increment as

$$dw_t = w_{t+dt} - w_t. \tag{4.171}$$

The statistical average of the Wiener process vanishes:

$$\langle dw_t \rangle = 0. \tag{4.172}$$

Moreover, the Wiener process is white in the sense that its increments at different times are independent from each other:

$$\langle dw_t dw_{t'} \rangle = 0 \ (t \neq t'). \tag{4.173}$$

Furthermore, the square of the Wiener increment is with unit probability equal to the increment of time:

$$(dw_t)^2 = dt. \tag{4.174}$$

Therefore, dw_t can be regarded as a quantity of the order of $\sqrt{dt}$. We, in fact, have the following equalities up to $o(dt)$:

$$dt^2 = 0, \tag{4.175}$$

$$dt dw_t = 0. \tag{4.176}$$

We now introduce stochastic integrals. The Itô integral is defined by

$$\int_{t=0}^{\tau} f_t \cdot dw_t := \lim_{N \to \infty} \sum_{n=0}^{N-1} f_{ndt}(w_{(n+1)dt} - w_{ndt}), \tag{4.177}$$

with $Ndt = \tau$ fixed. Meanwhile, the Stratonovich integral is defined by

$$\int_{t=0}^{\tau} f_t \circ dw_t := \lim_{N \to \infty} \sum_{n=0}^{N-1} \frac{f_{ndt} + f_{(n+1)dt}}{2}(w_{(n+1)dt} - w_{ndt}). \tag{4.178}$$

Remarkably, these two integrals differ from each other in contrast to the Riemann integral. For example, from

$$\begin{aligned}
\int_{t=0}^{\tau} w_t \cdot dw_t &= \lim_{N\to\infty}\sum_{n=0}^{N-1} w_{ndt}(w_{(n+1)dt} - w_{ndt}) \\
&= \lim_{N\to\infty}\sum_{n=0}^{N-1} w_{ndt}w_{(n+1)dt} - \lim_{N\to\infty}\sum_{n=0}^{N-1} w_{ndt}w_{ndt} \\
&= \lim_{N\to\infty}\sum_{n=0}^{N-1} w_{ndt}w_{(n+1)dt} - \lim_{N\to\infty}\sum_{n=0}^{N-2} w_{(n+1)dt}w_{(n+1)dt} - w_0^2 \\
&= w_\tau^2 - \lim_{N\to\infty}\sum_{n=0}^{N-2}(w_{(n+1)dt} - w_{ndt})w_{(n+1)dt} - w_0^2 \\
&= w_\tau^2 - \lim_{N\to\infty}\sum_{n=0}^{N-2}[(w_{(n+1)dt} - w_{ndt})w_{ndt} + dt] - w_0^2 \\
&= w_\tau^2 - w_0^2 - \tau - \int_{t=0}^{\tau} w_t \cdot dw_t,
\end{aligned} \tag{4.179}$$

the Itô integral is evaluated as

$$\int_{t=0}^{\tau} w_t \cdot dw_t = \frac{w_\tau^2 - w_0^2 - \tau}{2}. \tag{4.180}$$

On the other hand, the Stratonovich integral is

$$\begin{aligned}
\int_{t=0}^{\tau} w_t \cdot dw_t &= \lim_{N\to\infty}\sum_{n=0}^{N-1}\frac{w_{ndt} + w_{(n+1)dt}}{2}(w_{(n+1)dt} - w_{ndt}) \\
&= \frac{1}{2}\lim_{N\to\infty}\sum_{n=0}^{N-1}(w_{(n+1)dt}^2 - w_{ndt}^2) \\
&= \frac{w_\tau^2 - w_0^2}{2}.
\end{aligned} \tag{4.181}$$

The Itô product has an important property called the nonanticipating property.[1] Since dw_t is independent from any quantities in previous times, we have

$$\langle f_t \cdot dw_t\rangle = \langle f_t\rangle\langle dw_t\rangle = 0, \tag{4.182}$$

as long as $f_t(x)$ satisfies the causality in the sense that it does not depend on quantities after time t.

[1] The Stratonovich product $f_t \circ dw_t$ does not have the nonanticipating property. Actually, it is averaged as $\langle f_t \circ dw_t\rangle = \left\langle \frac{f_t + f_{t+dt}}{2} \cdot dw_t\right\rangle = \frac{1}{2}\langle f_{t+dt} \cdot dw_t\rangle$, and f_{t+dt} is, in general, dependent on dw_t.

A.2 Rescaling of Variables

For simplicity, we below use the rescaled variables defined by

$$x = \tilde{x}\sqrt{\frac{k_B T^{\text{eff}}}{m}}, \tag{4.183}$$

$$v = \tilde{v}\sqrt{\frac{k_B T^{\text{eff}}}{m}}, \tag{4.184}$$

$$f_t(x) = \tilde{f}_t(\tilde{x})\sqrt{mk_B T}, \tag{4.185}$$

$$Q_t = \tilde{Q}_t k_B T^{\text{eff}}, \tag{4.186}$$

$$V_t(x) = \tilde{V}_t(\tilde{x}) k_B T^{\text{eff}}, \tag{4.187}$$

where the tildes indicate that the accompanying quantities are rescaled.

By using these rescaled variables, the underdamped Langevin dynamics read

$$d\tilde{x}_t = \tilde{v}_t dt, \tag{4.188}$$

$$d\tilde{v}_t = \tilde{f}_t(\tilde{x}_t)dt + \sum_\mu \left(-\frac{\tilde{\gamma}^\mu}{\varepsilon}\tilde{v}_t dt + \sqrt{\frac{2\tilde{\gamma}^\mu \tilde{T}^\mu}{\varepsilon}} dw_t^\mu \right), \tag{4.189}$$

$$\delta \tilde{Q}_t^\mu = \frac{1}{\varepsilon}\tilde{\gamma}^\mu \tilde{v}_t^2 - \sqrt{\frac{2\tilde{\gamma}^\mu \tilde{T}^\mu}{\varepsilon}} \tilde{v}_t \circ dw_t^\mu \tag{4.190}$$

$$= \frac{1}{\varepsilon}\tilde{\gamma}^\mu (\tilde{v}_t^2 - N\tilde{T}^\mu) - \sqrt{\frac{2\tilde{\gamma}^\mu \tilde{T}^\mu}{\varepsilon}} \tilde{v}_t \cdot dw_t^\mu, \tag{4.191}$$

where the conversion of the Stratonovich product into the Itô product is performed as

$$\begin{aligned} \tilde{v}_t \circ dw_t^\mu &= \frac{\tilde{v}_{t+dt} + \tilde{v}_t}{2} \cdot dw_t^\mu \\ &= \tilde{v}_t \cdot dw_t^\mu + \frac{1}{2} d\tilde{v}_t \cdot dw_t^\mu \\ &= \tilde{v}_t \cdot dw_t^\mu + N\sqrt{\frac{\tilde{\gamma}^\mu \tilde{T}^\mu}{2\varepsilon}} dt. \end{aligned} \tag{4.192}$$

Meanwhile, the overdamped Langevin dynamics with a single reservoir reduce to

$$d\tilde{x}_t = \varepsilon \tilde{f}_t(\tilde{x}_t)dt + \sqrt{2\varepsilon} dw_t, \tag{4.193}$$

$$\begin{aligned} \delta \tilde{Q}_t &= \varepsilon \tilde{f}_t(\tilde{x}_t)^2 dt + \sqrt{2\varepsilon} \tilde{f}_t(\tilde{x}_t) \circ dw_t \\ &= \varepsilon[\tilde{f}_t(\tilde{x}_t)^2 + (\partial_{\tilde{x}} \cdot \tilde{f}_t(\tilde{x}_t))]dt + \sqrt{2\varepsilon} \tilde{f}_t(\tilde{x}_t) \cdot dw_t. \end{aligned} \tag{4.194}$$

A.3 Derivation of the Time-Evolution Equation of the Generating Function

We here derive the time evolution of the generating function in the underdamped Langevin dynamics. Let us consider the quantity defined by

$$\delta\tilde{X}_t^i = \tilde{a}_t^{i(0)}(\tilde{x}_t, \tilde{v}_t)dt + \frac{1}{\varepsilon}\tilde{a}_t^{i(-1)}(\tilde{x}_t, \tilde{v}_t)dt + \sum_\mu \sqrt{\frac{2\tilde{\gamma}^\mu \tilde{T}^\mu}{\varepsilon}}\tilde{b}_t^{i\mu}(\tilde{x}_t, \tilde{v}_t)\cdot dw_t^\mu, \tag{4.195}$$

and its generating function

$$G_{\tilde{X},t}(\tilde{x}, \tilde{v}, \{\Lambda^i\}) := \left\langle \delta(\tilde{x}_t - \tilde{x})\delta(\tilde{v}_t - \tilde{v})\exp\left[\sum_i \Lambda^i \tilde{X}_t^i\right]\right\rangle. \tag{4.196}$$

To this aim, we consider a stochastic function

$$\mathscr{F}_t(\tilde{x}, \tilde{v}, \{\Lambda^i\}) = \delta(\tilde{x}_t - \tilde{x})\delta(\tilde{v}_t - \tilde{v})\exp\left[\sum_i \Lambda^i \tilde{X}_t^i\right]. \tag{4.197}$$

The infinitesimal increment of $\mathscr{F}_t(\tilde{x}, \tilde{v}, \{\Lambda^i\})$ is given by

$$\begin{aligned} d\mathscr{F}_t(\tilde{x}, \tilde{v}, \{\Lambda^i\}) &= (\partial_{\tilde{x}_t}\mathscr{F}_t)\cdot d\tilde{x}_t + (\partial_{\tilde{v}_t}\mathscr{F}_t)\cdot d\tilde{v}_t + \frac{1}{2}(\partial_{\tilde{v}_t}^2\mathscr{F}_t)(d\tilde{v}_t)^2 \\ &\quad + \sum_i \Lambda^i\mathscr{F}_t\delta\tilde{X}_t^i + \frac{1}{2}\sum_{i,j}\Lambda^i\Lambda^j\mathscr{F}_t\delta\tilde{X}_t^i\delta\tilde{X}_t^j + \sum_i \Lambda^i(\partial_{\tilde{v}_t}\mathscr{F}_t)\cdot d\tilde{v}_t\delta\tilde{X}_t^i \\ &= -\partial_{\tilde{x}}\cdot(\mathscr{F}_t d\tilde{x}_t) - \partial_{\tilde{v}}\cdot(\mathscr{F}_t d\tilde{v}_t) + \frac{1}{2}\partial_{\tilde{v}}^2(\mathscr{F}_t(d\tilde{v}_t)^2) \\ &\quad + \sum_i \Lambda^i\mathscr{F}_t\delta\tilde{X}_t^i + \frac{1}{2}\sum_{i,j}\Lambda^i\Lambda^j\mathscr{F}_t\delta\tilde{X}_t^i\delta\tilde{X}_t^j - \sum_i \Lambda^i\partial_{\tilde{v}}\cdot(\mathscr{F}_t d\tilde{v}_t\delta\tilde{X}_t^i). \end{aligned} \tag{4.198}$$

Note that we have to keep the quadratic terms of $d\tilde{v}_t$ and $\delta\tilde{X}_t$ since they have contributions of $\mathscr{O}(dt)$ from terms proportional to dw_t^2. We take the statistical average of this equation to obtain the time evolution of the generating function. The first term on the right-hand side of Eq. (4.198) is averaged as

$$\begin{aligned} \langle\partial_{\tilde{x}}\cdot(\mathscr{F}_t d\tilde{x}_t)\rangle &= \partial_{\tilde{x}}\cdot\langle\mathscr{F}_t v_t dt\rangle \\ &= \partial_{\tilde{x}}\cdot(vG_{\tilde{X},t})dt. \end{aligned} \tag{4.199}$$

The average of the second term on the right-hand side of Eq. (4.198) is calculated as

$$
\begin{aligned}
\langle \partial_{\tilde{v}} \cdot (\mathscr{F}_t d\tilde{v}_t) \rangle &= \partial_{\tilde{v}} \cdot \langle \mathscr{F}_t d\tilde{v}_t \rangle \\
&= \partial_{\tilde{v}} \cdot \left\langle \mathscr{F}_t \cdot \left[\tilde{f}_t(\tilde{x}_t)dt + \sum_\mu \left(-\frac{\tilde{\gamma}^\mu}{\varepsilon} \tilde{v}_t dt + \sqrt{\frac{2\tilde{\gamma}^\mu T^\mu}{\varepsilon}} dw_t^\mu \right) \right] \right\rangle \\
&= \tilde{f}_t(\tilde{x}) \cdot (\partial_{\tilde{v}} G_{\tilde{X},t})dt - \frac{1}{\varepsilon} \partial_{\tilde{v}} \cdot (v G_{\tilde{X},t})dt,
\end{aligned} \tag{4.200}
$$

where we use the nonanticipating property of the Itô product. The average of the third term on the right-hand side of Eq. (4.198) is given by

$$
\begin{aligned}
\left\langle \frac{1}{2} \partial_{\tilde{v}}^2 (\mathscr{F}_t (d\tilde{v}_t)^2) \right\rangle &= \frac{1}{2} \partial_{\tilde{v}}^2 \langle \mathscr{F}_t (d\tilde{v}^2) \rangle \\
&= \frac{1}{2} \partial_{\tilde{v}}^2 \left\langle \mathscr{F}_t \left(\sum_\mu \sqrt{\frac{2\tilde{\gamma}^\mu \tilde{T}^\mu}{\varepsilon}} dw_t^\mu \sum_\nu \sqrt{\frac{2\tilde{\gamma}^\mu \tilde{T}^\nu}{\varepsilon}} dw_t^\nu + o(dt) \right) \right\rangle \\
&= \frac{1}{\varepsilon} (\partial_{\tilde{v}}^2 G_{\tilde{X},t})dt + o(dt).
\end{aligned} \tag{4.201}
$$

In a similar manner, the averages of the other terms of Eq. (4.198) can be evaluated. Consequently, we obtain the time-evolution operator of the generating function as

$$
\mathscr{L}_{\tilde{X},t}(\{\Lambda^i\}) = \mathscr{L}_{\tilde{X},t}^{(0)}(\{\Lambda^i\}) + \frac{1}{\varepsilon} \mathscr{L}_{\tilde{X},t}^{(-1)}(\{\Lambda^i\}), \tag{4.202}
$$

where we define

$$
\begin{aligned}
\mathscr{L}_{\tilde{X},t}^{(0)}(\{\Lambda^i\}) &= -\tilde{v} \cdot \partial_{\tilde{x}} - \tilde{f}_t(\tilde{x}) \partial_{\tilde{v}} + \sum_i \Lambda^i \tilde{a}^{(0)}(\tilde{x}, \tilde{v}), \qquad (4.203) \\
\mathscr{L}_{\tilde{X},t}^{(-1)}(\{\Lambda^i\}) &= \partial_{\tilde{v}}^2 + \partial_{\tilde{v}} \cdot \tilde{v} + \sum_i \Lambda^i \tilde{a}^{(-1)}(\tilde{x}, \tilde{v}) \\
&\quad + \sum_{i,j,\mu} \Lambda^i \Lambda^j \gamma^\mu T^\mu \tilde{b}^{i\mu}(\tilde{x}, \tilde{v}) \cdot \tilde{b}^{j\mu}(\tilde{x}, \tilde{v}) \\
&\quad -2 \sum_{i,\mu} \Lambda^i \gamma^\mu T^\mu \partial_{\tilde{v}} \cdot \tilde{b}^{i\mu}(\tilde{x}, \tilde{v}). \qquad (4.204)
\end{aligned}
$$

By returning to the original scale, we obtain Eqs. (4.51) and (4.52).

A.4 Derivation of the Underdamped Fluctuation Theorems

We here derive the fluctuation theorem for the underdamped Langevin dynamics.

For later reference, we consider the time-evolution operator of the heat-generating function:

$$\mathscr{L}_{\tilde{Q},t}(\{\Lambda^\mu\}) = \mathscr{L}_t^{(0)} + \frac{1}{\varepsilon}\mathscr{L}_{\tilde{Q}}^{(-1)}(\{\Lambda^\mu\}), \tag{4.205}$$

where

$$\mathscr{L}_t^{(0)} = -\tilde{v}\cdot\partial_{\tilde{x}} - \tilde{f}_t(\tilde{x})\cdot\partial_{\tilde{v}}, \tag{4.206}$$

$$\mathscr{L}_{\tilde{Q}}^{(-1)}(\{\Lambda^\mu\}) = \partial_{\tilde{v}}^2 + (1+B)\partial_{\tilde{v}}\cdot\tilde{v} + B\tilde{v}\cdot\partial_{\tilde{v}} + A\tilde{v}^2, \tag{4.207}$$

$$A(\{\Lambda^\mu\}) = \sum_\mu \Lambda^\mu\tilde{\gamma}^\mu(1+\Lambda^\mu\tilde{T}^\mu), \tag{4.208}$$

$$B(\{\Lambda^\mu\}) = \sum_\mu \Lambda^\mu\tilde{\gamma}^\mu\tilde{T}^\mu. \tag{4.209}$$

By noting

$$A(\{-\Lambda^\mu - \tilde{\beta}^\mu\}) = A(\{\Lambda^\mu\}), \tag{4.210}$$

$$B(\{-\Lambda^\mu - \tilde{\beta}^\mu\}) = -B(\{\Lambda^\mu\}) - 1, \tag{4.211}$$

we obtain

$$\mathscr{L}_{\tilde{Q}}^{(-1)\dagger}(\{-\Lambda^\mu - \tilde{\beta}^\mu\}) = \mathscr{L}_{\tilde{Q}}^{(-1)}(\{\Lambda^\mu\}). \tag{4.212}$$

On the other hand, the zeroth-order term satisfies

$$\mathscr{L}_t^{(0)\ddagger} = \mathscr{L}_t^{(0)}. \tag{4.213}$$

Recall that $\ddagger$ indicates the velocity inversion and the Hermitian conjugation at the same time. Since $\mathscr{L}^{(-1)}$ is even with respect to the velocity, the time-evolution operator has the symmetry

$$\mathscr{L}_{\tilde{Q},t}^{\ddagger}(\{-\Lambda^\mu - \tilde{\beta}^\mu\}) = \mathscr{L}_{\tilde{Q},t}(\{\Lambda^\mu\}). \tag{4.214}$$

In particular, when we consider the entropy production of the reservoirs defined by

$$\Delta s^{\mathrm{r}} = \sum_\mu \frac{Q^\mu}{k_{\mathrm{B}}T^\mu}, \tag{4.215}$$

its time-evolution operator

$$\mathscr{L}_{\Delta s^{\mathrm{r}},t}(\Lambda) = \mathscr{L}_{\tilde{Q},t}(\{\Lambda^\mu = \tilde{\beta}^\mu\Lambda\}) \tag{4.216}$$

takes over the symmetry

$$\mathscr{L}_{\Delta s^{\mathrm{r}},t}^{\ddagger}(-\Lambda - 1) = \mathscr{L}_{\Delta s^{\mathrm{r}},t}(\Lambda). \tag{4.217}$$

A.4.1 Underdamped Finite-Time Integral Fluctuation Theorem for the Total Entropy Production

We here consider the total entropy production

$$\Delta s^{\text{tot}} = \Delta s + \Delta s^{\text{r}} \tag{4.218}$$

and its generating function

$$G_{\Delta s^{\text{tot}},t}(\tilde{x}, \tilde{v}, \Lambda) = \left\langle \delta(\tilde{x}_t - \tilde{x})\delta(\tilde{v}_t - \tilde{v}) \exp[\Lambda \Delta s^{\text{tot}}] \right\rangle. \tag{4.219}$$

Although we can directly derive the time evolution of this generating function, it is convenient to utilize the time evolution of the reservoir entropy production. By noting

$$\partial_t G_{\Delta s^{\text{tot}},t}(\tilde{x}, \tilde{v}, \Lambda) = \partial_t (\left\langle \delta(\tilde{x}_t - \tilde{x})\delta(\tilde{v}_t - \tilde{v}) \exp[\Lambda \Delta s^{\text{r}}] \right\rangle P_t(\tilde{x}, \tilde{v})^{-\Lambda}), \tag{4.220}$$

we obtain the time-evolution operator as

$$\mathscr{L}_{\Delta s^{\text{tot}},t}(\Lambda) = P_t(\tilde{x}, \tilde{v})^{-\Lambda} \mathscr{L}_{\Delta s^{\text{r}},t}(\Lambda) P_t(\tilde{x}, \tilde{v})^{\Lambda} - \Lambda(\partial_t \ln P_t(\tilde{x}, \tilde{v})). \tag{4.221}$$

From the symmetry of Eq. (4.217), we obtain

$$\mathscr{L}_{\Delta s^{\text{tot}},t}(\Lambda) = P_t(\tilde{x}, \tilde{v})^{-\Lambda} \mathscr{L}^{\ddagger}_{\Delta s^{\text{r}},t}(-\Lambda - 1) P_t(\tilde{x}, \tilde{v})^{\Lambda} - \Lambda(\partial_t \ln P_t(\tilde{x}, \tilde{v})). \tag{4.222}$$

When we set $\Lambda = -1$, this equation reduces to

$$\mathscr{L}_{\Delta s^{\text{tot}},t}(-1) = P_t(\tilde{x}, \tilde{v}) \mathscr{L}^{\text{K}\ddagger}_t P_t(\tilde{x}, \tilde{v})^{-1} + (\partial_t \ln P_t(\tilde{x}, \tilde{v})), \tag{4.223}$$

where $\mathscr{L}^{\text{K}}$ is the Kramers operator (4.53). Therefore, we obtain

$$\begin{aligned}
&\partial_t [P_t(\tilde{x}, \tilde{v})^{-1} G_{\Delta s^{\text{tot}},t}(\tilde{x}, \tilde{v}, -1)] \\
&= P_t(\tilde{x}, \tilde{v})^{-1} [\mathscr{L}_{\Delta s^{\text{tot}},t}(-1) - (\partial_t \ln P_t(\tilde{x}, \tilde{v}))] G_{\Delta s^{\text{tot}},t}(\tilde{x}, \tilde{v}, -1) \\
&= \mathscr{L}^{\text{K}\ddagger}_t [P_t(\tilde{x}, \tilde{v})^{-1} G_{\Delta s^{\text{tot}},t}(\tilde{x}, \tilde{v}, -1)].
\end{aligned} \tag{4.224}$$

When $P_t(\tilde{x}, \tilde{v}) \neq 0$, this equation can formally be solved as

$$\begin{aligned}
G_{\Delta s^{\text{tot}},\tau}(\tilde{x}, \tilde{v}, -1) &= P_\tau(\tilde{x}, \tilde{v}) \mathbf{T}_t \exp\left[\int_0^\tau dt\ \mathscr{L}^{\text{K}\ddagger}_t\right] P_0(\tilde{x}, \tilde{v})^{-1} G_{\Delta s^{\text{tot}},0}(\tilde{x}, \tilde{v}, -1) \\
&= P_\tau(\tilde{x}, \tilde{v}) \mathbf{T}_t \exp\left[\int_0^\tau dt\ \mathscr{L}^{\text{K}\ddagger}_t\right],
\end{aligned} \tag{4.225}$$

where $\mathbf{T}_t$ is the time-ordering operator with respect to time t. Since $P_0(\tilde{x}, \tilde{v}) \neq 0$ implies $P_t(\tilde{x}, \tilde{v}) \neq 0$, Eq. (4.225) holds for $P_0(\tilde{x}, \tilde{v}) \neq 0$. Meanwhile, when

$P_0(\tilde{x}, \tilde{v}) = 0$, the generating function vanishes:

$$G_{\Delta s^{\mathrm{tot}},\tau}(\tilde{x}, \tilde{v}, -1) = \mathbf{T}_t \exp\left[\int_0^\tau dt\ \mathscr{L}_{\Delta s^{\mathrm{tot}},t}(-1)\right] P_0(\tilde{x}, \tilde{v}) = 0. \tag{4.226}$$

By integrating the generating function over $\tilde{x}$ and $\tilde{v}$, we obtain

$$\begin{aligned}\langle e^{-\Delta s^{\mathrm{tot}}}\rangle &= \int_{P_0(\tilde{x},\tilde{v})\neq 0} d\tilde{x}d\tilde{v}\ G_{\Delta s^{\mathrm{tot}},\tau}(\tilde{x}, \tilde{v}, -1)\\ &= \int_{P_0(\tilde{x},\tilde{v})\neq 0} d\tilde{x}d\tilde{v}\ P_\tau(\tilde{x}, \tilde{v})\mathbf{T}_t \exp\left[\int_0^\tau dt\ \mathscr{L}_t^{\mathrm{K}\ddagger}\right]\\ &= \int_{P_0(\tilde{x},-\tilde{v})\neq 0} d\tilde{x}d\tilde{v}\ \bar{\mathbf{T}}_t \exp\left[\int_0^\tau dt\ \mathscr{L}_t^{\mathrm{K}}\right] P_\tau(\tilde{x}, -\tilde{v}),\end{aligned} \tag{4.227}$$

where $\bar{\mathbf{T}}_t$ is the anti-time-ordering operator. By defining the inverted time $\bar{t} := \tau - t$, we obtain

$$\langle e^{-\Delta s^{\mathrm{tot}}}\rangle = \int_{P_0(\tilde{x},-\tilde{v})\neq 0} d\tilde{x}d\tilde{v}\ \mathbf{T}_{\bar{t}} \exp\left[\int_0^\tau d\bar{t}\ \bar{\mathscr{L}}_{\bar{t}}^{\mathrm{K}}\right] \bar{P}_0(\tilde{x}, \tilde{v}), \tag{4.228}$$

where we define the Kramers operator in the time-reversed process $\bar{\mathscr{L}}_{\bar{t}}^{\mathrm{K}} := \mathscr{L}_{\tau-\bar{t}}^{\mathrm{K}}$ and the initial state of the time-reversed process as $\bar{P}_0(\tilde{x}, \tilde{v}) := P_\tau(\tilde{x}, -\tilde{v})$. Since the integrand is nothing but the probability distribution function at time τ in the time-reversed process, we obtain the integral fluctuation theorem

$$\langle e^{-\Delta s^{\mathrm{tot}}}\rangle = \int_{P_0(\tilde{x},-\tilde{v})\neq 0} d\tilde{x}d\tilde{v}\ \bar{P}_\tau(\tilde{x}, \tilde{v}) \tag{4.229}$$

$$= 1 - \lambda, \tag{4.230}$$

where we define the degree of absolute irreversibility by

$$\lambda = \int_{P_0(\tilde{x},-\tilde{v})=0} d\tilde{x}d\tilde{v}\ \bar{P}_\tau(\tilde{x}, \tilde{v}). \tag{4.231}$$

Thus, Eq. (4.63) has been proven.

A.4.2 Underdamped Finite-Time Fluctuation Theorems for the Irreversible Entropy Production

The irreversible entropy production is defined by

$$\Delta_{\mathrm{i}} s = \Delta(-\ln P^{\mathrm{eq}}) + \Delta s^{\mathrm{r}} \tag{4.232}$$

$$= \frac{W - \Delta F - \sum_{\mu} \eta^{\mu} Q^{\mu}}{k_{\mathrm{B}} T^0}. \tag{4.233}$$

By a procedure similar to the total entropy production, we obtain the time-evolution operator

$$\mathscr{L}_{\Delta_{\mathrm{i}} s, t}(\Lambda) = P_t^{\mathrm{eq}}(\tilde{x}, \tilde{v})^{-\Lambda} \mathscr{L}_{\Delta s^{\mathrm{r}}, t}(\Lambda) P_t^{\mathrm{eq}}(\tilde{x}, \tilde{v})^{\Lambda} - \Lambda(\partial_t \ln P_t^{\mathrm{eq}}(\tilde{x}, \tilde{v})). \tag{4.234}$$

From the symmetry of Eq. (4.217), this equation reduces to

$$\mathscr{L}_{\Delta_{\mathrm{i}} s, t}(\Lambda) = P_t^{\mathrm{eq}}(\tilde{x}, \tilde{v})^{-\Lambda} \mathscr{L}^{\ddagger}_{\Delta s^{\mathrm{r}}, t}(-\Lambda - 1) P_t^{\mathrm{eq}}(\tilde{x}, \tilde{v})^{\Lambda} - \Lambda(\partial_t \ln P_t^{\mathrm{eq}}(\tilde{x}, \tilde{v})). \tag{4.235}$$

Meanwhile, from Eq. (4.234), we obtain

$$\begin{aligned}\mathscr{L}^{\ddagger}_{\Delta_{\mathrm{i}} s, t}(-\Lambda - 1) = {} & P_t^{\mathrm{eq}}(\tilde{x}, \tilde{v})^{-\Lambda-1} \mathscr{L}^{\ddagger}_{\Delta s^{\mathrm{r}}, t}(-\Lambda - 1) P_t^{\mathrm{eq}}(\tilde{x}, \tilde{v})^{\Lambda+1} \\ & +(\Lambda + 1)(\partial_t \ln P_t^{\mathrm{eq}}(\tilde{x}, \tilde{v})),\end{aligned} \tag{4.236}$$

where we use the symmetry of the equilibrium state $P_t^{\mathrm{eq}}(\tilde{x}, \tilde{v}) = P_t^{\mathrm{eq}}(\tilde{x}, -\tilde{v})$. Therefore, we find the symmetry

$$\mathscr{L}_{\Delta_{\mathrm{i}} s, t}(\Lambda) = P_t^{\mathrm{eq}}(\tilde{x}, \tilde{v}) \mathscr{L}^{\ddagger TR}_{\Delta_{\mathrm{i}} s, t}(-\Lambda - 1) P_t^{\mathrm{eq}}(\tilde{x}, \tilde{v})^{-1} + (\partial_t \ln P_t^{\mathrm{eq}}(\tilde{x}, \tilde{v})), \tag{4.237}$$

where the superscript TR indicates the local time reversal ($\partial_t \to -\partial_t$). Since

$$\begin{aligned}\partial_t [P_t^{\mathrm{eq}}(\tilde{x}, \tilde{v})^{-1} G_{\Delta_{\mathrm{i}} s, t}(\tilde{x}, \tilde{v}, \Lambda)] &= P_t^{\mathrm{eq}}(\tilde{x}, \tilde{v})^{-1} [\mathscr{L}_{\Delta_{\mathrm{i}} s, t}(\Lambda) - (\partial_t \ln P_t^{\mathrm{eq}}(\tilde{x}, \tilde{v}))] G_{\Delta_{\mathrm{i}} s, t}(\tilde{x}, \tilde{v}, \Lambda) \\ &= \mathscr{L}^{\ddagger TR}_{\Delta_{\mathrm{i}} s, t}(-\Lambda - 1)[P_t^{\mathrm{eq}}(\tilde{x}, \tilde{v})^{-1} G_{\Delta_{\mathrm{i}} s, t}(\tilde{x}, \tilde{v}, \Lambda)],\end{aligned} \tag{4.238}$$

the generating function can formally be written as

$$G_{\Delta_{\mathrm{i}} s, \tau}(\tilde{x}, \tilde{v}, \Lambda) = P_\tau^{\mathrm{eq}}(\tilde{x}, \tilde{v}) \mathbf{T}_t \exp\left[\int_0^\tau dt\, \mathscr{L}^{\ddagger TR}_{\Delta_{\mathrm{i}} s, t}(-\Lambda - 1)\right]. \tag{4.239}$$

By integrating this over $\tilde{x}$ and $\tilde{v}$, we obtain

$$
\begin{aligned}
\int d\tilde{x}d\tilde{v}\; G_{\Delta_i s,\tau}(\tilde{x},\tilde{v},\Lambda) &= \int d\tilde{x}d\tilde{v}\; P_\tau^{\mathrm{eq}}(\tilde{x},\tilde{v})\mathbf{T}_t \exp\left[\int_0^\tau dt\; \mathscr{L}_{\Delta_i s,t}^{\ddagger TR}(-\Lambda-1)\right] \\
&= \int d\tilde{x}d\tilde{v}\bar{\mathbf{T}}_t \exp\left[\int_0^\tau dt\; \mathscr{L}_{\Delta_i s,t}^{TR}(-\Lambda-1)\right] P_\tau^{\mathrm{eq}}(\tilde{x},\tilde{v}) \\
&= \int d\tilde{x}d\tilde{v}\; \mathbf{T}_{\bar{t}} \exp\left[\int_0^\tau d\bar{t}\; \bar{\mathscr{L}}_{\Delta_i s,\bar{t}}(-\Lambda-1)\right] \bar{P}_0^{\mathrm{eq}}(\tilde{x},\tilde{v}) \\
&= \int d\tilde{x}d\tilde{v}\; \bar{G}_{\Delta_i s,\tau}(\tilde{x},\tilde{v},-\Lambda-1),
\end{aligned} \tag{4.240}
$$

where $\bar{G}$ is the generating function in the time-reversed process in which the state starts from the equilibrium state $\bar{P}_0^{\mathrm{eq}} = P_\tau^{\mathrm{eq}}$ and the systematic force is applied in the time-reversed manner. In terms of the probability distribution function of the irreversible entropy production $P(\Delta_i s)$ and its time-reversed one $\bar{P}(\Delta_i s)$, this equality is rewritten as

$$
\begin{aligned}
\int d\Delta_i s\; P(\Delta_i s)\exp[\Lambda\Delta_i s] &= \int d\Delta_i s\; \bar{P}(\Delta_i s)\exp[(-\Lambda-1)\Delta_i s] \\
&= \int d\Delta_i s\; \bar{P}(-\Delta_i s)e^{\Delta_i s}\exp[\Lambda\Delta_i s].
\end{aligned} \tag{4.241}
$$

Therefore, we obtain the finite-time detailed fluctuation theorem (4.69), i.e.,

$$
\frac{\bar{P}(-\Delta_i s)}{P(\Delta_i s)} = e^{-\Delta_i s}, \tag{4.242}
$$

and the integral fluctuation theorem (4.70), i.e.,

$$
\langle e^{-\Delta_i s}\rangle = 1. \tag{4.243}
$$

A.4.3 Underdamped Asymptotic Steady-State Fluctuation Theorem for Heat

As we have already seen, the heat-generating function has the symmetry (4.72), i.e.,

$$
\mathscr{L}_{\tilde{Q},t}^{\ddagger}(\{-\Lambda^\mu - \tilde{\beta}^\mu\}) = \mathscr{L}_{\tilde{Q},t}(\{\Lambda^\mu\}). \tag{4.244}
$$

Since the adjoint operation and the velocity inversion do not change the eigenvalues of an operator, the operator $\mathscr{L}_{\tilde{Q},t}(\{-\Lambda^\mu - \tilde{\beta}^\mu\})$ has the same eigenvalues as $\mathscr{L}_{\tilde{Q},t}(\{\Lambda^\mu\})$. In particular, the largest eigenvalues of these operators, which dominate the long-time behavior of the generating functions, are equal. As a consequence, when we consider a situation in which the systematic force $f_t(x)$ is constant in time, the generating function has asymptotically the symmetry

$$\mathscr{G}_{\tilde{Q},t}(\{-\Lambda^{\mu}-\tilde{\beta}^{\mu}\}) \asymp \mathscr{G}_{\tilde{Q},t}(\{\Lambda^{\mu}\}), \tag{4.245}$$

where the symbol $\asymp$ indicates the asymptotic equality, or to be precise

$$\lim_{t\to\infty}\frac{1}{t}\ln\frac{\mathscr{G}_{\tilde{Q},t}(\{-\Lambda^{\mu}-\tilde{\beta}^{\mu}\})}{\mathscr{G}_{\tilde{Q},t}(\{\Lambda^{\mu}\})}=0. \tag{4.246}$$

Therefore, one might expect that the steady-state fluctuation theorem holds for the joint probability distribution function of heat. However, it is not always the case because of the van Zon-Cohen singularity [17, 18].

The large deviation function of the probability distribution function is defined by

$$I(\{\tilde{q}^{\mu}\}) := -\lim_{t\to\infty}\frac{1}{t}\ln P_t(\{\tilde{Q}^{\mu}=\tilde{q}^{\mu}t\}), \tag{4.247}$$

or equivalently

$$P_t(\{\tilde{Q}^{\mu}=\tilde{q}^{\mu}t\}) \asymp \exp[-tI(\{\tilde{q}^{\mu}\})]. \tag{4.248}$$

For ordinary situations, the large deviation function can be evaluated by the Legendre transformation of the largest eigenvalue α of the time-evolution operator of the generating function as

$$I(\{\tilde{q}^{\mu}\}) = \max_{\{\Lambda^{\mu}\}}\left[\sum_{\mu}\Lambda^{\mu}\tilde{q}^{\mu}-\alpha(\{\Lambda^{\mu}\})\right]. \tag{4.249}$$

Therefore, from the symmetry

$$\alpha(\{-\Lambda^{\mu}-\tilde{\beta}^{\mu}\})=\alpha(\{\Lambda^{\mu}\}), \tag{4.250}$$

we obtain

$$I(\{\tilde{q}^{\mu}\}) = \max_{\{\Lambda^{\mu}\}}\left[\sum_{\mu}\Lambda^{\mu}\tilde{q}^{\mu}-\alpha(\{-\Lambda^{\mu}-\tilde{\beta}^{\mu}\})\right] \tag{4.251}$$

$$= \max_{\{\Lambda^{\mu}\}}\left[\sum_{\mu}(-\Lambda^{\mu}-\tilde{\beta}^{\mu})\tilde{q}^{\mu}-\alpha(\{\Lambda^{\mu}\})\right] \tag{4.252}$$

$$= I(\{-\tilde{q}^{\mu}\})-\sum_{\mu}\tilde{\beta}^{\mu}\tilde{q}^{\mu}. \tag{4.253}$$

Consequently, the steady-state fluctuation theorem holds:

$$\lim_{t\to\infty}\frac{1}{t}\ln\frac{P_t(\{\tilde{Q}^{\mu}=\tilde{q}^{\mu}t\})}{P_t(\{\tilde{Q}^{\mu}=-\tilde{q}^{\mu}t\})}=I(\{-\tilde{q}^{\mu}\})-I(\{\tilde{q}^{\mu}\})=\sum_{\mu}\tilde{\beta}^{\mu}\tilde{q}^{\mu}, \tag{4.254}$$

or, in the original variables,

$$\lim_{t\to\infty}\frac{1}{t}\ln\frac{P_t(\{Q^\mu = q^\mu t\})}{P_t(\{Q^\mu = -q^\mu t\})} = \sum_\mu \frac{q^\mu}{k_\mathrm{B}T^\mu}. \tag{4.255}$$

However, the Legendre transformation (4.249) does not always connect the large deviation function and the largest eigenvalue of the time-evolution operator of the generating function. In such cases, the steady-state fluctuation theorem does not hold. Equation (4.249) is proven by the saddle-point approximation of the inverse Fourier-Laplace transform of the generating function. When the generating function

$$\mathcal{G}_t(\{\Lambda^\mu\}) \asymp g(\{\Lambda^\mu\})\exp[t\alpha(\{\Lambda^\mu\})] \tag{4.256}$$

has singularities in the sense that $g(\{\Lambda^\mu\})$ has a pole or a branch cut along the path of the saddle-point approximation, Eq. (4.249) no longer holds. As a result, the steady-state fluctuation theorem (4.255) ceases to be valid, although the generating function has the symmetry (4.245). This breakdown of the steady-state fluctuation theorem is known as the van Zon-Cohen singularity [17–19], which sometimes occurs in nonisothermal systems [20, 21].

A.5 Overdamped Theory for the Entropy Productions

We have shown that the overdamped heat generating function

$$G^{\mathrm{od}}_{\tilde{Q},t}(\tilde{x},\{\Lambda^\mu\}) = \left\langle \delta(\tilde{x}_t - \tilde{x})\exp\left[\sum_\mu \Lambda^\mu \tilde{Q}^{\mu,\mathrm{od}}\right]\right\rangle \tag{4.257}$$

evolves in time by the operator

$$\mathcal{L}^{\mathrm{od}}_{\tilde{Q},t}(\{\Lambda^\mu\}) = \frac{\varepsilon}{R^2}\left[\partial_{\tilde{x}}^2 - \kappa\partial_{\tilde{x}}\cdot\tilde{f}_t(\tilde{x}) + \rho\tilde{f}_t(\tilde{x})\cdot\partial_{\tilde{x}} + A\tilde{f}_t(\tilde{x})^2\right], \tag{4.258}$$

with

$$\kappa(\{\Lambda^\mu\}) = \frac{1+2B+R}{2}, \quad (4.259)$$

$$\rho(\{\Lambda^\mu\}) = \frac{-1-2B+R}{2}, \quad (4.260)$$

$$R(\{\Lambda^\mu\}) = \sqrt{(1+2B)^2 - 4A}, \quad (4.261)$$

$$A(\{\Lambda^\mu\}) = \sum_\mu \Lambda^\mu \tilde{\gamma}^\mu (1 + \Lambda^\mu \tilde{T}^\mu), \quad (4.262)$$

$$B(\{\Lambda^\mu\}) = \sum_\mu \Lambda^\mu \tilde{\gamma}^\mu \tilde{T}^\mu. \quad (4.263)$$

We can define the overdamped entropy production of the reservoir by

$$\Delta s^{\mathrm{r,od}} := \sum_\mu \frac{Q^{\mu,\mathrm{od}}}{k_{\mathrm{B}} T^\mu}. \quad (4.264)$$

Then, the time-evolution operator of its generating function is given by

$$\mathscr{L}^{\mathrm{od}}_{\Delta s^{\mathrm{r}},t}(\Lambda) = \mathscr{L}^{\mathrm{od}}_{\tilde{Q},t}(\{\Lambda^\mu = \tilde{\beta}^\mu \Lambda\}), \quad (4.265)$$

or more explicitly

$$\mathscr{L}^{\mathrm{od}}_{\Delta s^{\mathrm{r}},t}(\Lambda) = \frac{\varepsilon}{\mathfrak{R}^2}\left[\partial_{\tilde{x}}^2 - \mathfrak{K}\partial_{\tilde{x}} \cdot \tilde{f}_t(\tilde{x}) + \mathfrak{P}\tilde{f}_t(\tilde{x}) \cdot \partial_{\tilde{x}} + \mathfrak{A}\tilde{f}(\tilde{x})^2\right], \quad (4.266)$$

with

$$\mathfrak{K}(\Lambda) = \frac{1+2\Lambda+\mathfrak{R}}{2}, \quad (4.267)$$

$$\mathfrak{P}(\Lambda) = \frac{-1-2\Lambda+\mathfrak{R}}{2}, \quad (4.268)$$

$$\mathfrak{R}(\Lambda) = \sqrt{(1+2\Lambda)^2 - 4\mathfrak{A}}, \quad (4.269)$$

$$\mathfrak{A}(\Lambda) = \Lambda(1+\Lambda)\sum_\mu \tilde{\gamma}^\mu \tilde{\beta}^\mu. \quad (4.270)$$

By using this operator, we here derive the overdamped theory for the total entropy production and the irreversible entropy production.

A.5.1 Overdamped Theory for the Total Entropy Production

The total entropy production in the overdamped approximation is given by

$$\Delta s^{\mathrm{tot,od}} = \Delta s^{\mathrm{od}} + \sum_{\mu} \frac{Q^{\mu,\mathrm{od}}}{k_{\mathrm{B}} T^{\mu}}. \tag{4.271}$$

Then, its overdamped generating function reads

$$G^{\mathrm{od}}_{\Delta s^{\mathrm{tot}},t}(\Lambda) = \langle \delta(\tilde{x}_t - \tilde{x}) \exp[\Lambda \Delta s^{\mathrm{tot,od}}] \rangle. \tag{4.272}$$

By a procedure similar to the one to derive Eq. (4.221), we obtain the time-evolution operator

$$\mathscr{L}^{\mathrm{od}}_{\Delta s^{\mathrm{tot}},t}(\Lambda) = P^{\mathrm{od}}_t(\tilde{x})^{-\Lambda} \mathscr{L}^{\mathrm{od}}_{\Delta s^{\mathrm{r}},t}(\Lambda) P^{\mathrm{od}}_t(\tilde{x})^{\Lambda} - \Lambda(\partial_t \ln P^{\mathrm{od}}_t(\tilde{x})). \tag{4.273}$$

Therefore, by inserting Eq. (4.266), we obtain the overdamped time-evolution operator

$$\begin{aligned}
&\mathscr{L}^{\mathrm{od}}_{\Delta s^{\mathrm{tot}},t}(\Lambda) \\
&= -\Lambda(\partial_t \ln P^{\mathrm{od}}_t(\tilde{x})) \\
&\quad + \frac{\varepsilon}{\mathfrak{R}^2} \Big[\partial^2_{\tilde{x}} + 2\Lambda(\partial_{\tilde{x}} \ln P^{\mathrm{od}}_t(\tilde{x})) \cdot \partial_{\tilde{x}} - \mathfrak{K} \partial_{\tilde{x}} \cdot \tilde{f}_t(\tilde{x}) + \mathfrak{P} \tilde{f}_t(\tilde{x}) \cdot \partial_{\tilde{x}} \\
&\quad + \Lambda(\partial^2_{\tilde{x}} \ln P^{\mathrm{od}}_t(\tilde{x})) + \Lambda^2 (\partial_{\tilde{x}} \ln P^{\mathrm{od}}_t(\tilde{x}))^2 \\
&\quad - \Lambda(1+2\Lambda) \tilde{f}_t(\tilde{x}) \cdot (\partial_{\tilde{x}} \ln P^{\mathrm{od}}(\tilde{x})) + \mathfrak{A} \tilde{f}_t(\tilde{x})^2 \Big].
\end{aligned} \tag{4.274}$$

Meanwhile, the contribution in the fast time scale is given by

$$\mathscr{G}^{v}_{\Delta s^{\mathrm{tot}},t}(\Lambda) = \exp\left[\frac{N(1-\mathfrak{R})t}{2\varepsilon} \right]. \tag{4.275}$$

When there is only one reservoir, from relations $\mathfrak{A} = \Lambda(1+\Lambda)$, $\mathfrak{R} = 1$, $\mathfrak{K} = 1 + \Lambda$ and $\mathfrak{P} = -\Lambda$, the time-evolution operator reduces to

$$\begin{aligned}
&\mathscr{L}^{\mathrm{od}}_{\Delta s^{\mathrm{tot}},t}(\Lambda) \\
&= -\Lambda(\partial_t \ln P^{\mathrm{od}}_t(\tilde{x})) \\
&\quad + \varepsilon \Big[\partial^2_{\tilde{x}} + 2\Lambda(\partial_{\tilde{x}} \ln P^{\mathrm{od}}_t(\tilde{x})) \cdot \partial_{\tilde{x}} - (1+\Lambda)\partial_{\tilde{x}} \cdot \tilde{f}_t(\tilde{x}) - \Lambda \tilde{f}_t(\tilde{x}) \cdot \partial_{\tilde{x}} \\
&\quad + \Lambda(\partial^2_{\tilde{x}} \ln P^{\mathrm{od}}_t(\tilde{x})) + \Lambda^2(\partial_{\tilde{x}} \ln P^{\mathrm{od}}_t(\tilde{x}))^2 \\
&\quad - \Lambda(1+2\Lambda)\tilde{f}_t(\tilde{x}) \cdot (\partial_{\tilde{x}} \ln P^{\mathrm{od}}(\tilde{x})) + \Lambda(1+\Lambda)\tilde{f}_t(\tilde{x})^2 \Big],
\end{aligned} \tag{4.276}$$

which coincides with the time-evolution operator (4.116) derived from the overdamped Langevin dynamics with a single reservoir for

$$\begin{aligned}\delta\Delta s^{\mathrm{tot,od}} &= -(\partial_t \ln P_t^{\mathrm{od}}(\tilde{x}))dt - (\partial_{\tilde{x}} \ln P_t^{\mathrm{od}}(\tilde{x})) \circ d\tilde{x}_t + \delta\tilde{Q}_t &(4.277)\\ &= -(\partial_t \ln P_t^{\mathrm{od}}(\tilde{x}))dt \\ &\quad +\varepsilon[\tilde{f}_t^2(\tilde{x})^2 + (\partial_{\tilde{x}} \cdot \tilde{f}_t(\tilde{x})) - \tilde{f}_t(\tilde{x}) \cdot (\partial_{\tilde{x}} \ln P_t^{\mathrm{od}}(\tilde{x})) - (\partial_{\tilde{x}}^2 \ln P_t^{\mathrm{od}}(\tilde{x}))] \\ &\quad +\sqrt{2\varepsilon}[\tilde{f}_t(\tilde{x}) - (\partial_{\tilde{x}} \ln P_t^{\mathrm{od}}(\tilde{x}))] \cdot dw_t. &(4.278)\end{aligned}$$

A.5.2 Overdamped Theory for the Irreversible Entropy Production

The overdamped irreversible entropy production is given by

$$\begin{aligned}\Delta_{\mathrm{i}} s^{\mathrm{od}} &= \Delta(-\ln P_t^{\mathrm{eq,od}}) + \sum_{\mu} \frac{Q^{\mu,\mathrm{od}}}{k_{\mathrm{B}} T^{\mu}} &(4.279)\\ &= \frac{\Delta V - \Delta F^{\mathrm{od}}}{k_{\mathrm{B}} T^0} + \sum_{\mu} \frac{Q^{\mu,\mathrm{od}}}{k_{\mathrm{B}} T^{\mu}} &(4.280)\\ &\simeq \frac{W - \Delta F^{\mathrm{od}} - \sum_{\mu} \eta^{\mu} Q^{\mu,\mathrm{od}}}{k_{\mathrm{B}} T^0}, &(4.281)\end{aligned}$$

where we use the first law of thermodynamics (4.158) to obtain the last line. Its overdamped generating function

$$G^{\mathrm{od}}_{\Delta s^{\mathrm{i}},t}(\Lambda) = \langle \delta(\tilde{x}_t - \tilde{x}) \exp[\Lambda \Delta_{\mathrm{i}} s^{\mathrm{od}}] \rangle \tag{4.282}$$

evolves by the operator

$$\mathscr{L}^{\mathrm{od}}_{\Delta_{\mathrm{i}} s,t}(\Lambda) = P_t^{\mathrm{eq,od}}(\tilde{x})^{-\Lambda} \mathscr{L}^{\mathrm{od}}_{\Delta s^{\mathrm{r}},t}(\Lambda) P_t^{\mathrm{eq,od}}(\tilde{x})^{\Lambda} - \Lambda(\partial_t \ln P_t^{\mathrm{eq,od}}(\tilde{x})). \tag{4.283}$$

Therefore, the explicit form of the time-evolution operator is given by

$$\begin{aligned}&\mathscr{L}^{\mathrm{od}}_{\Delta_{\mathrm{i}} s,t}(\Lambda) \\ &= \Lambda(\partial_t \tilde{V}_t(\tilde{x})) - \Lambda(\partial_t \tilde{F}_t^{\mathrm{od}}) \\ &\quad + \frac{\varepsilon}{\mathfrak{R}^2}\Big[\partial_{\tilde{x}}^2 - \partial_{\tilde{x}} \cdot [\mathfrak{K}\tilde{f}_t(\tilde{x}) - \Lambda\tilde{\beta}^0 \tilde{f}_t^{\mathrm{c}}(\tilde{x})] + [\mathfrak{P}\tilde{f}_t(\tilde{x}) + \Lambda\tilde{\beta}^0 \tilde{f}_t^{\mathrm{c}}(\tilde{x})] \cdot \partial_{\tilde{x}} \\ &\quad + \Lambda^2(\tilde{\beta}^0)^2 \tilde{f}_t^{\mathrm{c}}(\tilde{x})^2 - \Lambda(1+2\Lambda)\tilde{\beta}^0 \tilde{f}_t(\tilde{x}) \cdot \tilde{f}_t^{\mathrm{c}}(\tilde{x}) + \mathfrak{A}\tilde{f}_t(\tilde{x})^2\Big]. &(4.284)\end{aligned}$$

The contribution from the fast time scale is given by

$$\mathscr{G}^{v}_{\Delta_{\mathrm{i}} s,t}(\Lambda) = \exp\left[\frac{N(1-\mathfrak{R})t}{2\varepsilon}\right]. \tag{4.285}$$

When there is only one reservoir, the time-evolution operator reduces to

$$\begin{aligned}
&\mathscr{L}^{\mathrm{od}}_{\Delta_{\mathrm{i}}s,t}(\Lambda)\\
&= \Lambda(\partial_t \tilde{V}_t(\tilde{x})) - \Lambda(\partial_t \tilde{F}^{\mathrm{od}}_t)\\
&\quad+\varepsilon\Big[\partial_{\tilde{x}}^2 - \partial_{\tilde{x}} \cdot [(1+\Lambda)\tilde{f}_t(\tilde{x}) - \Lambda \tilde{f}^{\mathrm{c}}_t(\tilde{x})] - \Lambda[\tilde{f}_t(\tilde{x}) - \tilde{f}^{\mathrm{c}}_t(\tilde{x})] \cdot \partial_{\tilde{x}}\\
&\quad+\Lambda^2 \tilde{f}^{\mathrm{c}}_t(\tilde{x})^2 - \Lambda(1+2\Lambda)\tilde{f}_t(\tilde{x}) \cdot \tilde{f}^{\mathrm{c}}_t(\tilde{x}) + \Lambda(1+\Lambda)\tilde{f}_t(\tilde{x})^2\Big] && (4.286)\\
&= \Lambda(\partial_t \tilde{V}_t(\tilde{x})) - \Lambda(\partial_t \tilde{F}^{\mathrm{od}}_t)\\
&\quad+\varepsilon\Big[\partial_{\tilde{x}}^2 - \partial_{\tilde{x}} \cdot [\tilde{f}_t(\tilde{x}) + \Lambda \tilde{f}^{\mathrm{nc}}_t(\tilde{x})] - \Lambda \tilde{f}^{\mathrm{nc}}_t(\tilde{x}) \cdot \partial_{\tilde{x}} + \Lambda^2 \tilde{f}^{\mathrm{nc}}_t(\tilde{x})^2 + \Lambda \tilde{f}_t(\tilde{x}) \cdot \tilde{f}^{\mathrm{nc}}_t(\tilde{x})\Big], && (4.287)
\end{aligned}$$

which is identical to the time-evolution operator (4.116) for the dissipated work

$$\begin{aligned}
\delta\Delta_{\mathrm{i}}s^{\mathrm{od}} &= (\partial_t \tilde{V}_t(\tilde{x}))dt + \tilde{f}^{\mathrm{nc}}_t(\tilde{x}_t) \circ d\tilde{x}_t - (\partial_t \tilde{F}^{\mathrm{od}}_t)dt && (4.288)\\
&= [(\partial_t \tilde{V}_t(\tilde{x}) - (\partial_t \tilde{F}^{\mathrm{od}}_t)]dt\\
&\quad+\varepsilon[\tilde{f}_t(\tilde{x}) \cdot \tilde{f}^{\mathrm{nc}}_t(\tilde{x}) + (\partial_{\tilde{x}} \tilde{f}^{\mathrm{nc}}_t(\tilde{x}))]dt + \sqrt{2\varepsilon}\, \tilde{f}^{\mathrm{nc}}_t(\tilde{x}) \cdot dw_t. && (4.289)
\end{aligned}$$

A.6 Derivation of the Overdamped Fluctuation Theorems

From the relations

$$A(\{-\Lambda^\mu - \tilde{\beta}^\mu\}) = A(\{\Lambda^\mu\}), \tag{4.290}$$
$$B(\{-\Lambda^\mu - \tilde{\beta}^\mu\}) = -B(\{\Lambda^\mu\}) - 1, \tag{4.291}$$

we obtain

$$R(\{\Lambda^\mu - \tilde{\beta}^\mu\}) = R(\{\Lambda^\mu\}), \tag{4.292}$$
$$\kappa(\{-\Lambda^\mu - \tilde{\beta}^\mu\}) = \rho(\{\Lambda^\mu\}), \tag{4.293}$$
$$\rho(\{-\Lambda^\mu - \beta^\mu\}) = \kappa(\{\Lambda^\mu\}). \tag{4.294}$$

Therefore, the overdamped time-evolution operator of the heat-generating function satisfies the symmetry of

$$\mathscr{L}^{\mathrm{od}\dagger}_{\tilde{Q},t}(\{-\Lambda^\mu - \tilde{\beta}^\mu\}) = \mathscr{L}^{\mathrm{od}}_{\tilde{Q},t}(\{\Lambda^\mu\}). \tag{4.295}$$

Hence, the time-evolution operator for the generating function of the reservoir entropy production takes over the symmetry

$$\mathscr{L}^{\mathrm{od}\dagger}_{\Delta s^{\mathrm{r}},t}(-\Lambda - 1) = \mathscr{L}^{\mathrm{od}}_{\Delta s^{\mathrm{r}},t}(\Lambda). \tag{4.296}$$

A.6.1 Overdamped Finite-Time Integral Fluctuation Theorem for the Total Entropy Production

From Eq. (4.273) and the symmetry (4.296), we obtain

$$\mathscr{L}^{\mathrm{od}}_{\Delta s^{\mathrm{tot}},t}(\Lambda) = P^{\mathrm{od}}_t(\tilde{x})^{-\Lambda}\mathscr{L}^{\mathrm{od}^\dagger}_{\Delta s^{\mathrm{r}},t}(-\Lambda-1)P^{\mathrm{od}}_t(\tilde{x})^{\Lambda} - \Lambda(\partial_t \ln P^{\mathrm{od}}_t(\tilde{x})). \quad (4.297)$$

When we set $\Lambda = -1$, this equation reduces to

$$\mathscr{L}^{\mathrm{od}}_{\Delta s^{\mathrm{tot}},t}(-1) = P^{\mathrm{od}}_t(\tilde{x})\mathscr{L}^{\mathrm{K}\dagger}_t P^{\mathrm{od}}_t(\tilde{x})^{-1} + (\partial_t \ln P^{\mathrm{od}}_t(\tilde{x})), \quad (4.298)$$

where $\mathscr{L}^{\mathrm{K}}$ is the Kramers operator. Therefore, the integral fluctuation theorem can be derived:

$$\begin{aligned}
\langle e^{-\Delta s^{\mathrm{tot,od}}}\rangle &= \int_{P^{\mathrm{od}}_0(\tilde{x})\neq 0} d\tilde{x}\; G^{\mathrm{od}}_{\Delta s^{\mathrm{tot}},\tau}(\tilde{x},-1) \\
&= \int_{P^{\mathrm{od}}_0(\tilde{x})\neq 0} d\tilde{x}\; P^{\mathrm{od}}_\tau(\tilde{x})\mathbf{T}_t \exp\left[\int_0^\tau dt\; \mathscr{L}^{\mathrm{FP}\dagger}_t\right] \\
&= \int_{P^{\mathrm{od}}_0(\tilde{x})\neq 0} d\tilde{x}\bar{\mathbf{T}}_t \exp\left[\int_0^\tau dt\; \mathscr{L}^{\mathrm{FP}}_t\right] P^{\mathrm{od}}_\tau(\tilde{x}) \\
&= \int_{P^{\mathrm{od}}_0(\tilde{x})\neq 0} d\tilde{x}\; \mathbf{T}_{\bar{t}} \exp\left[\int_0^\tau d\bar{t}\; \mathscr{L}^{\bar{\mathrm{FP}}}_{\bar{t}}\right] \bar{P}^{\mathrm{od}}_0(\tilde{x}) \\
&= \int_{P^{\mathrm{od}}_0(\tilde{x})\neq 0} d\tilde{x}\; \bar{P}^{\mathrm{od}}_\tau(\tilde{x}) \\
&= 1-\lambda,
\end{aligned} \quad (4.299)$$

where the degree of absolute irreversibility is given by

$$\lambda := \int_{P^{\mathrm{od}}_0(\tilde{x})=0} d\tilde{x}\; \bar{P}^{\mathrm{od}}_\tau(\tilde{x}). \quad (4.300)$$

Thus, Eq. (4.162) has been proven.

A.6.2 Overdamped Finite-Time Fluctuation Theorems for the Irreversible Entropy Production

From Eq. (4.283) and the symmetry (4.296), the time-evolution operator for the irreversible entropy production satisfies

$$\begin{aligned}
\mathscr{L}^{\mathrm{od}}_{\Delta_{\mathrm{i}} s,t}(\Lambda) = P^{\mathrm{eq,od}}_t(\tilde{x})^{-\Lambda}\mathscr{L}^{\mathrm{od}^\dagger}_{\Delta s^{\mathrm{r}},t}(-\Lambda-1)P^{\mathrm{eq,od}}_t(\tilde{x})^{\Lambda} \\
-\Lambda(\partial_t \ln P^{\mathrm{eq,od}}_t(\tilde{x})).
\end{aligned} \quad (4.301)$$

Meanwhile, from Eq. (4.283), we obtain

$$\mathscr{L}^{\mathrm{od}}_{\Delta_{\mathrm{i}}s,t}(-\Lambda-1) = P_t^{\mathrm{eq,od}}(\tilde{x})^{\Lambda+1}\mathscr{L}^{\mathrm{od}}_{\Delta s^{\mathrm{r}},t}(-\Lambda-1)P_t^{\mathrm{eq,od}}(\tilde{x})^{-\Lambda-1} \\ +(\Lambda+1)(\partial_t \ln P_t^{\mathrm{eq,od}}(\tilde{x})). \tag{4.302}$$

By taking the conjugate, we acquire

$$\mathscr{L}^{\mathrm{od}\dagger}_{\Delta_{\mathrm{i}}s,t}(-\Lambda-1) = P_t^{\mathrm{eq,od}}(\tilde{x})^{-\Lambda-1}\mathscr{L}^{\mathrm{od}\dagger}_{\Delta s^{\mathrm{r}},t}(-\Lambda-1)P_t^{\mathrm{eq,od}}(\tilde{x})^{\Lambda+1} \\ +(\Lambda+1)(\partial_t \ln P_t^{\mathrm{eq,od}}(\tilde{x})). \tag{4.303}$$

Hence, the time-evolution operator for the irreversible entropy production has the symmetry

$$\mathscr{L}^{\mathrm{od}}_{\Delta_{\mathrm{i}}s,t}(\Lambda) = P_t^{\mathrm{eq,od}}(\tilde{x})\mathscr{L}^{\mathrm{od}\dagger TR}_{\Delta_{\mathrm{i}}s,t}(-\Lambda-1)P_t^{\mathrm{eq,od}}(\tilde{x})^{-1} \\ +(\partial_t \ln P_t^{\mathrm{eq,od}}(\tilde{x})). \tag{4.304}$$

Therefore, the symmetry of the generating function is derived

$$\begin{aligned}
\mathscr{G}^{\mathrm{od}}_{\Delta_{\mathrm{i}}s,\tau}(\Lambda) &= \int d\tilde{x}\; G^{\mathrm{od}}_{\Delta_{\mathrm{i}}s,\tau}(\tilde{x},\Lambda) \\
&= \int d\tilde{x}\; P_\tau^{\mathrm{eq,od}}(\tilde{x})\mathbf{T}_t \exp\left[\int_0^\tau dt\; \mathscr{L}^{\mathrm{od}\dagger TR}_{\Delta_{\mathrm{i}}s,t}(-\Lambda-1)\right] \\
&= \int d\tilde{x}\bar{\mathbf{T}}_t \exp\left[\int_0^\tau dt\; \mathscr{L}^{\mathrm{od}TR}_{\Delta_{\mathrm{i}}s,t}(-\Lambda-1)\right] P_\tau^{\mathrm{eq,od}}(\tilde{x}) \\
&= \int d\tilde{x}\; \mathbf{T}_{\bar{t}} \exp\left[\int_0^\tau d\bar{t}\; \bar{\mathscr{L}}^{\mathrm{od}}_{\Delta_{\mathrm{i}}s,\bar{t}}(-\Lambda-1)\right] \bar{P}_0^{\mathrm{eq,od}}(\tilde{x}) \\
&= \int d\tilde{x}\; \bar{G}^{\mathrm{od}}_{\Delta_{\mathrm{i}}s,\tau}(\tilde{x},-\Lambda-1) \\
&= \mathscr{G}^{\mathrm{od}}_{\Delta_{\mathrm{i}}s,\tau}(-\Lambda-1),
\end{aligned} \tag{4.305}$$

which indicates the detailed fluctuation theorem (4.167), i.e.,

$$\frac{\bar{P}(-\Delta_{\mathrm{i}}s^{\mathrm{od}})}{P(\Delta_{\mathrm{i}}s^{\mathrm{od}})} = e^{-\Delta_{\mathrm{i}}s^{\mathrm{od}}} \tag{4.306}$$

and the integral fluctuation theorem (4.168), i.e.,

$$\langle e^{-\Delta_{\mathrm{i}}s^{\mathrm{od}}}\rangle = 1. \tag{4.307}$$

A.6.3 Overdamped Asymptotic Steady-State Fluctuation Theorem for Heat

Equation (4.295) is nothing but Eq. (4.170) and indicates the symmetry

$$\lim_{t\to\infty} \frac{1}{t} \ln \frac{\mathscr{G}^{\text{od}}_{\tilde{Q},t}(\{-\Lambda^\mu - \tilde{\beta}^\mu\})}{\mathscr{G}^{\text{od}}_{\tilde{Q},t}(\{\Lambda^\mu\})} = 0. \tag{4.308}$$

In the absence of the van Zon-Cohen singularity, this symmetry is equivalent to the steady-state fluctuation theorem

$$\lim_{t\to\infty} \frac{1}{t} \ln \frac{P_t(\{\tilde{Q}^{\mu,\text{od}} = \tilde{q}^{\mu,\text{od}}t\})}{P_t(\{\tilde{Q}^{\mu,\text{od}} = -\tilde{q}^{,\text{od}}t\})} = \sum_\mu \tilde{\beta}^\mu \tilde{q}^{\mu,\text{od}}. \tag{4.309}$$

References

1. Murashita Y, Esposito M (2016) Phys Rev E 94:062148
2. Feynman RP, Leighton RB, Sands M (1963) The Feynman lectures on physics, vol 1. Addison-Wesley Publishing Company
3. Parrondo JMR, Español P (1996) Am J Phys 64:1125
4. Sekimoto K (1997) J Phys Soc Jpn 66:1234
5. Sekimoto K (2010) Stochastic enegetics. Springer
6. Fogedby HC, Imparato A (2012) J Stat Mech Theor Exp P04005
7. Dotsenko V, Maciołek A, Vasilyev O, Oshanin G (2013) Phys Rev E 87:062130
8. Chun HM, Noh JD (2015) Phys Rev E 91:052128
9. Magnasco MO, Stolovitzky G (1998) J Stat Phys 93:615
10. Gomez-Marin A, Sancho JM (2005) Phys Rev E 71:021101
11. Van den Broeck C, Esposito M (2010) Phys Rev E 82:011144
12. Esposito M, Van den Broeck C (2010) Phys Rev E 82:011143
13. Bulnes Cuetara G, Esposito M, Imparato A (2014) Phys Rev E 89:052119
14. Matsuo M, Sasa S (2000) Physica A 276:188
15. Celani A, Bo S, Eichhorn R, Aurell E (2012) Phys Rev Lett 109:260603
16. Bo S, Celani A (2014) J Stat Phys 154:1325
17. van Zon R, Cohen EGD (2003) Phys Rev Lett 91:110601
18. van Zon R, Cohen EGD (2004) Phys Rev E 69:056121
19. Noh JD, Park JM (2012) Phys Rev Lett 108:240603
20. Visco P (2006) J Stat Mech Theor Exp P06006
21. Sabhapandit S (2012) Phys Rev E 85:021108

Chapter 5
Gibbs Paradox and the Fluctuation Theorems

In this chapter, we show an intimate relation between the Gibbs paradox and the fluctuation theorem with absolute irreversibility. First of all, we outline some historical discussions on the gas mixing given by Gibbs and show that the Gibbs paradox is rooted in the foundations of thermodynamics and statistical mechanics. Then, following van Kampen, we classify the Gibbs paradox into three distinct aspects: the consistency within thermodynamics, the consistency within statistical mechanics, and the inter-theoretical consistency between thermodynamics and statistical mechanics. Next, we review the conventional resolutions of these issues. Then, we point out that the resolution for the last aspect is restricted to the thermodynamic limit since it is based on extensivity of the thermodynamic entropy. Finally, we demonstrate that the fluctuation theorem with absolute irreversibility takes the place of extensivity to settle the issue in small thermodynamic systems where the thermodynamic entropy is non-extensive.

Most of the contents of this chapter is based on the published letter [1]. Although the author has discussed the relation between the Gibbs paradox and absolute irreversibility in his master thesis [2], the discussion was restricted to the mixing of non-interacting gases. In contrast, the discussion in this thesis applies to the mixing of interacting gases as long as interactions do not break additivity.

5.1 Historical Discussions

The Gibbs paradox collectively refers to problems concerning the particle-number dependence of the entropy in thermodynamics and statistical mechanics. Here we review some historical discussions on the Gibbs paradox. Gibbs considered mixing of identical and different gases and pointed out that their thermodynamic entropy productions differ even when their dynamics are identical [3]. He argued that this counterintuitive behavior is rooted in the definition of thermodynamic states [3]. Later on, he revisited the problem of gas mixing in his renowned textbook on statistical

Y. Murashita, *Fluctuation Theorems under Divergent Entropy Production and their Applications for Fundamental Problems in Statistical Physics*, Springer Theses, https://doi.org/10.1007/978-981-16-8638-2_5

mechanics [4]. Therein he introduced two phases of statistical mechanics, i.e., the generic phase and the specific phase [4]. Then, he argued that only the generic phase leads to the correct thermodynamic entropy production upon gas mixing and we must therefore employ this phase [4]. The problem of gas mixing was later revisited from a viewpoint of the calculation of chemical potentials based on statistical mechanics [5].

5.1.1 Original Gibbs Paradox

In his paper [3, pp. 227–229], Gibbs drafted a section entitled "Considerations relating to the Increase of Entropy due to the Mixture of Gases by Diffusion." Gibbs compared the entropy production in mixing of identical gases (Fig. 5.1a) and different gases (Fig. 5.1b). Let us consider mixing of two different gases at a constant temperature and pressure. Suppose that each of them is in either side of an equally separated bipartite box and the partition is removed to mix the gases. When the gases are identical, the thermodynamic entropy production should vanish. In contrast, the entropy production upon different-gas mixing is calculated to be $2Nk_B \ln 2$.

Gibbs stressed that the extensive increase of the entropy upon different-gas mixing is independent of the kinds of gases as long as they are not the same. Even when the difference of two different gases becomes infinitesimal, the entropy production stays extensive and constant. In particular, we may imagine two different gases that have absolutely identical properties while they exist as gases, although they interact differently with some other substances. In this case, the dynamical behavior in the gas mixing of these gases is identical in its minutest details to that in the identical-gas mixing. Nevertheless, the entropy increases by the extensive amount. This fact seems paradoxical and "In such respects, entropy stands strongly contrasted with energy" [3, p. 229].

To understand this property of the entropy, we have to bear in mind the following considerations. When we refer to the entropy increase upon mixing of different

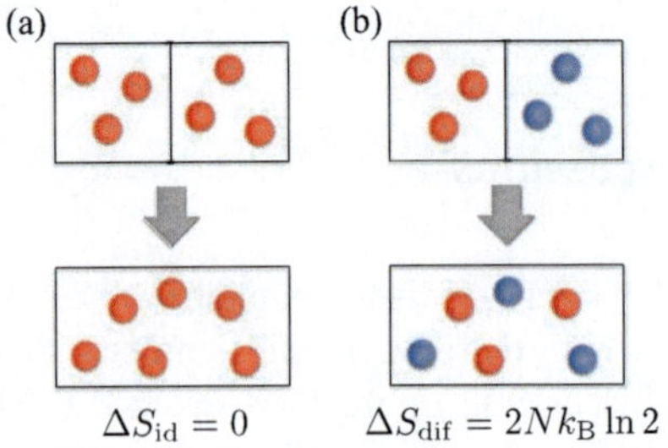

Fig. 5.1 Gas mixing and the Gibbs paradox. **a** Mixing of identical gases. The entropy production is assumed to be zero. **b** Mixing of different gases. An extensive amount of the entropy is produced. Even if the difference between two gases is infinitesimal, the entropy production stays constant and extensive. This behavior of the entropy seems paradoxical

gases, we mean that the mixed gas can be separated into the original state by means of external work corresponding to the increased amount of entropy. However, when we refer to no entropy production upon mixing of identical gases, we do not mean that the mixed gas can be separated again without work. Rather, the separation is impossible because we cannot identify any difference between gas particles from one side and the other. Therefore, when we say the state returns to the original state, we do not mean that the particles return to their original positions, but mean that its sensible properties return to those of the original state. In this way, "It is to states of systems thus incompletely defined that the problems of thermodynamics relate" [3, p. 228]. Thus, Gibbs ascribed the difference of the entropy increase of the two gas mixings not to their dynamics but to the definition of thermodynamic states. His insightful observation has later been clarified by van Kampen [6] and Jaynes [7].

5.1.2 Discussion in the Renowned Textbook by Gibbs

In his renowned textbook on statistical mechanics [4, Chap. XV], Gibbs revisited the problem of gas mixing in a slightly different context. Therein he considered a gas composed of several kinds and variation in the numbers of particles. He introduced two distinct phases: the generic phase and the specific phase. The difference of these phases is whether or not we regard the phase, in which certain identical particles are exchanged their positions with one another, as identical to the original phase. In the generic phase these phases are supposed to be identical, while in the specific phase they are not. Gibbs argued that we should decide which phase we use "in accordance with the requirements of practical convenience in the discussion of the problems with which we are engaged" [4, p. 188].

Let the numbers of particles of the different kinds be denoted by $n_1, n_2, \ldots, n_h$. Then, the number of specific phases corresponding to one generic phase is $n_1!n_2!\ldots n_h!$. Then, the probability density of a generic phase is the sum of those of the corresponding specific phases. When these probabilities are the same, the probability density of the generic phase is that of each of the specific phases multiplied by $n_1!n_2!\ldots n_h!$. Gibbs noted that the statistical equilibrium in the specific phase should also be that in the generic phase, but not vice versa. This is because different specific phases corresponding to the same generic phase have different probabilities if exchanges of identical particles result in different probabilities.

Then, Gibbs constructed the grand canonical ensemble. In the specific phase, the probability density of the grand canonical ensemble is assumed to be

$$\frac{e^{-\beta(E-\Omega-\mu_1 n_1-\cdots-\mu_h n_h)}}{n_1!\ldots n_h!}, \tag{5.1}$$

where E is the energy of the system, Ω is the grand potential, and $\mu_1, \ldots, \mu_h$ are the chemical potentials in the modern terms. Since this value is the same for all the specific phases corresponding to one generic phase, the probability density in the generic phase is

$$e^{-\beta(E-\Omega-\mu_1 n_1-\cdots-\mu_h n_h)}. \tag{5.2}$$

Let us prepare two systems in the grand canonical ensembles with the same temperature and chemical potentials and combine them. The probability densities in the two systems can be written as

$$\frac{e^{-\beta(E'-\Omega'-\mu_1 n_1'-\cdots-\mu_h n_h')}}{n_1'!\ldots n_h'!}, \tag{5.3}$$

$$\frac{e^{-\beta(E''-\Omega''-\mu_1 n_1''-\cdots-\mu_h n_h'')}}{n_1''!\ldots n_h''!} \tag{5.4}$$

in the specific phase, respectively, and

$$e^{-\beta(E'-\Omega'-\mu_1 n_1'-\cdots-\mu_h n_h')}, \tag{5.5}$$

$$e^{-\beta(E''-\Omega''-\mu_1 n_1''-\cdots-\mu_h n_h'')} \tag{5.6}$$

in the generic phase, respectively. The probability density in the combined system can be calculated as the product of the probability density of each system as

$$\frac{e^{-\beta(E'''-\Omega'''-\mu_1 n_1'''-\cdots-\mu_h n_h''')}}{n_1'!\ldots n_h'!n_1''!\ldots n_h''!} \tag{5.7}$$

in the specific phase, where $E''' = E' + E''$, $\Omega''' = \Omega' + \Omega''$, $n_1''' = n_1' + n_1''$, etc., and

$$e^{-\beta(E'''-\Omega'''-\mu_1 n_1'''-\cdots-\mu_h n_h''')} \tag{5.8}$$

in the generic phase. If the probability density were equally distributed for the specific phases corresponding to the same generic phase, it should be

$$\frac{e^{-\beta(E'''-\Omega'''-\mu_1 n_1'''-\cdots-\mu_h n_h''')}}{n_1'''!\ldots n_h'''!}, \tag{5.9}$$

which is obviously different from Eq. (5.7). The reason of this discrepancy is that the specific phases obtained by interchange of particles in two subsystems have vanishing probability. As long as the subsystems do not physically exchange particles, Eqs. (5.7) and (5.8) represent statistical equilibria in the specific and generic phases, respectively. Now, we let the subsystems exchange particles to mix them. Then, as far as the interaction between particles is negligibly small, the generic ensemble (5.8) approximately remains in the statistical equilibrium, but the specific ensemble (5.7) does not since the equilibrium is given by Eq. (5.9).

Finally, Gibbs turned to consider the method of calculating thermodynamic quantities on the basis of a grand canonical ensemble. The entropy can be calculated both in the specific and generic phases. However, in mixing of identical gases, we assume no change in the entropy. Therefore, the entropies calculated before and after mixing should be the same. Hence, we have to invoke the generic phase where the equilibrium ensemble does not change upon mixing. "When the number of particles in a system is to be treated as variable, the average index of probability for phases generically defined corresponds to entropy[1]" [4, p. xviii]. In this way, Gibbs demonstrated that we should choose the statistical mechanics that is consistent with the thermodynamic requirement.

5.1.3 Later Discussions in View of Statistical Mechanics

In the 1910s, the problem of the particle-number dependence of the entropy was revisited for calculating chemical potentials on the basis of statistical mechanics (see, e.g., Refs. [8–10]). To determine the particle-number dependence of the entropy, we have to invoke a process in which particle numbers reversibly change. To achieve this kind of reversible processes, one can consider association and dissociation in gas mixture. Ehrenfest and Trkal [5] argued that one can derive the factor $n_1! \dots n_h!$ by considering association and dissociation of molecules in view of statistical mechanics in the full phase space of atoms. Thus, the particle-number dependence of the entropy can be determined in statistical mechanics.

5.2 Classification and Resolutions of the Gibbs Paradox

In his article titled "The Gibbs Paradox," van Kampen analyzed this "time-worn subject" [6]. Here, following him, we classify it into three distinct subjects: consistency within thermodynamics (GP-I), consistency within statistical mechanics (GP-II), and inter-theoretical consistency between thermodynamics and statistical mechanics (GP-III) as schematically illustrated in Fig. 5.2. We review a conventional resolution for each of these aspects.

5.2.1 Consistency Within Thermodynamics (GP-I)

We here consider the consistency within thermodynamics. As reviewed in Sect. 5.1.1, Gibbs ascribed the difference of the entropy productions in two gas-mixing processes to the definition of thermodynamic states [3]. Let us elucidate his idea in line with van Kampen [6, Sect. 2-4].

In macroscopic thermodynamics, we define the thermodynamic entropy by the Clausius equality. We change the state attached to an environment so slowly that

[1] Here the thermodynamic entropy is implied.

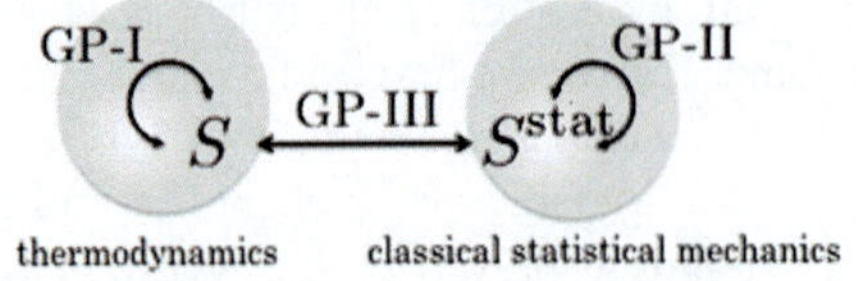

Fig. 5.2 Three aspects of the Gibbs paradox. The problems of the Gibbs paradox can be classified into three aspects, namely, the consistency within thermodynamics (GP-I), within statistical mechanics (GP-II), and between thermodynamics and statistical mechanics (GP-III). Reprinted figure with permission from Ref. [1]. Copyright 2017 by the American Physical Society

the state can stay in equilibrium. In this reversible process, we accumulate the heat exchange divided by the temperature to define the entropy difference. In the case of a classical ideal monoatomic gas, the entropy is defined as

$$S(T, P) = \frac{5}{2} N k_{\mathrm{B}} \ln T - N k_{\mathrm{B}} \ln P + C. \tag{5.10}$$

By this method, we can define "only entropy differences between states that can be connected by a reversible change" [6, p. 305]. Therefore, the constant C may depend on any quantities other than T or P; in particular, C depends on the particle number N.

To determine the N dependence of the entropy, one should connect states with different N reversibly. To this aim, we consider vessels containing identical gases with the same T, P, N. We can achieve a reversible process by opening up a channel between the two vessels. Therefore, the entropy of the initial state $2S(T, P, N)$ is equal to that of the final state $S(T, P, 2N)$, leading to the conclusion that $S(T, P, N)$ is proportional to N. We have here assumed that the entropy satisfies additivity. Thus, for an ideal gas, we obtain

$$S(T, P, N) = \frac{5}{2} N k_{\mathrm{B}} \ln T - N k_{\mathrm{B}} \ln P + cN. \tag{5.11}$$

If two vessels contain different gases, opening the channel is no longer a reversible process. Instead of this process, we have to use the process with semipermeable walls to obtain the entropy of the mixed gas as

$$S(T, P, N_A, N_B) = \frac{5}{2}(N_A + N_B) k_{\mathrm{B}} \ln T - N_A k_{\mathrm{B}} \ln P_A - N_B k_{\mathrm{B}} \ln P_B + c_A N_A + c_B N_B, \tag{5.12}$$

where

$$P_A = \frac{N_A}{N_A + N_B} P, \quad P_B = \frac{N_B}{N_A + N_B} P. \tag{5.13}$$

The fact

$$S(T, P, N_A + N_B) \neq S(T, P, N_A, N_B)|_{c_A = c_B} \tag{5.14}$$

indicates the Gibbs paradox.

As we see, the difference of the entropy productions arises from the fact that we utilize two different processes to define the entropy, and these processes are exclusive. Suppose that A and B are so similar that an experimenter has no way to distinguish them. Then, she/he does not have the process with the semi-permeable walls and opening the channel looks reversible for her/him. Moreover, Eq. (5.11) does not lead her/him to any inconsistent results within her/his capability. Therefore, actual increase in the entropy has no physical meaning for her/him. Let us conclude this subsection by citing van Kampen [6, pp. 306–307]:

> Thus, whether such a process is reversible or not depends on how discriminating the observer is. The expression for the entropy [...] depends on whether or not he is able or willing to distinguish between the molecules A and B. This is a paradox only for those who attach more physical reality to the entropy than is implied by its definition.

Incidentally, Jaynes revisited the thermodynamic aspect of the Gibbs paradox and concluded that the entropy has "anthropomorphic" nature in that it depends on "human information" [7, Sect. 5].

5.2.2 Consistency Within Statistical Mechanics (GP-II)

As we see in Sect. 5.1.3, the factorial factor can be deduced within statistical mechanics. Here we review how we can do this in line with van Kampen [6, Sect. 5-8].

We consider a system attached to a heat reservoir. Then, the probability in phase space is given by the canonical distribution as

$$W(q, p) = \text{const. } e^{-\beta H^*(q,p)}, \tag{5.15}$$

where H^* is the Hamiltonian of the system. Then, let us assume that the system can be divided into two subsystems with particle numbers N and N', where the sum of them is fixed: $N + N' = N^*$. The total Hamiltonian H^* can be written as $H_N + H'_{N^*-N}$ for each value of N as long as one can ignore interactions between the subsystems. To calculate the probability for the first subsystem, one should choose N particles out of N^* and then integrate out the degrees of freedom in the second subsystem. Therefore, the probability for the first subsystem is

$$W(N, q, p) = \text{const.} \binom{N^*}{N} e^{-\beta H_N(q,p)} \int e^{-\beta H'_{N'}(q',p')} dq'dp'. \tag{5.16}$$

By taking the limits of $N^* \to \infty$ and $V^* \to \infty$ with N^*/V^* fixed, this formula reduces to

$$W(N, q, p) = \text{const. } \frac{z^N}{N!} e^{-\beta H_N(q,p)}, \tag{5.17}$$

where z is a function of β. "The $N!$ arises from the computation of phase space volume according to the original rules without any additional postulate, either classical or quantummechanical" [6, p. 309].

5.2.3 *Consistency Between Thermodynamics and Statistical Mechanics (GP-III)*

As the last aspect of the Gibbs paradox, we consider the consistency between thermodynamics and statistical mechanics. As we discussed in Sect. 5.1.2, Gibbs chose the statistical mechanics that is consistent with the thermodynamic requirement on gas mixing. In fact, we always face the same problem when we statistical-mechanically calculate the thermodynamic entropy. The discussion below is partly based on van Kampen [6, Sect. 9] and Jaynes [7, Sect. 9].

We denote by S the thermodynamic entropy, whose change is defined by the Clausius equality. Then, we empirically know that it satisfies the thermodynamic relation as

$$TdS(E, V) = dE + PdV, \tag{5.18}$$

where E is the internal energy of the system. On the other hand, we can show that the entropy S^{stat} calculated on the basis of the canonical ensemble and the standard method in statistical mechanics satisfies the same relation as

$$TdS^{\mathrm{stat}}(E, V) = dE + PdV. \tag{5.19}$$

This is why Gibbs identified S^{stat} with S in his textbook [4, Chap. IV]. Since Eqs. (5.18) and (5.19) are formulae with no variation of the particle number, we have ambiguity in the relation between S and S^{stat} as

$$S(E, V, N) = S^{\mathrm{stat}}(E, V, N) + k_{\mathrm{B}} f(N). \tag{5.20}$$

Now, the problem is how to determine the ambiguity function $f(N)$. In particular, when we use statistical mechanics with the specific phase, $f(N)$ should be determined to be $\ln(1/N!)$ up to a constant per particle.

In his article, van Kampen [6, Sect. 9] dismissed this problem by stating that $f(N)$ is set to zero by convention (for classical statistical mechanics based on the generic phase or quantum statistical mechanics). However, Jaynes [7, Sect. 7-9], inspired by Pauli's analysis on the phenomenological entropy [11], argued that $f(N)$ can be determined by invoking extensivity. By way of illustration, let us consider ideal gas in the statistical mechanics based on the specific phase. The canonical distribution

$$\frac{1}{(2\pi m k_{\mathrm{B}} T)^{3N/2} V^N} \exp\left[-\sum_{n=1}^{N} \frac{p_n^2}{2m k_{\mathrm{B}} T}\right] \tag{5.21}$$

leads to a non-extensive entropy in statistical mechanics as

$$S^{\text{stat}}(T, V, N) = Nk_{\text{B}} \left[\frac{3}{2} \ln(2\pi m k_{\text{B}} T) + N \ln V + \frac{3}{2} \right]. \tag{5.22}$$

Then, if we require that the thermodynamic entropy S be extensive as

$$S(T, qV, qN) = qS(T, V, N), \ {}^{\forall}q > 0, \tag{5.23}$$

we obtain a functional equation

$$f(qN) = qf(N) - qN \ln q. \tag{5.24}$$

By solving this equation, we obtain

$$f(N) = Nf(1) - N \ln N, \tag{5.25}$$

where the first term represents the internal entropy of a particle and the second term shows the nontrivial dependence on N. We note that $f(1)$ still depends on the kind of particles. Remarkably, the latter term approaches $\ln(1/N!)$ in the thermodynamic limit because of the Stirling formula. Thus, we can determine the particle-number dependence of the thermodynamic entropy by requiring its extensivity. Moreover, the choice of the ambiguity function $f(N)$ in Eq. (5.25) is equivalent to extensivity:

$$[\text{thermodynamics with extensivity}] \Leftrightarrow f(N) = Nf(1) - N \ln N, \tag{5.26}$$

provided that we start from statistical mechanics based on the specific phase. We note that the same logic applies to such interacting cases as the van der Waals gas. In this way, extensivity in thermodynamics fixes the relation between thermodynamics and statistical mechanics.

5.2.4 Does Quantum Mechanics Resolve the Issues?

We here discuss the so-called quantum resolution of the Gibbs paradox. Many renowned textbooks (e.g., see Refs. [12–17]) argue that quantum mechanics resolves the Gibbs paradox. If we assumed the axiom that the thermodynamic entropy should be nothing but the entropy calculated by quantum statistical mechanics, the quantum resolution would explain the factorial factor and resolve the Gibbs paradox. However, this standpoint sometimes fails to explain phenomena in a mesoscopic regime. Moreover, it has a logical flaw from a viewpoint of thermodynamics as an operational theory.

5.2.4.1 Failure in the Mesoscopic Regime

By way of illustration, let us consider a system consisting of colloidal particles. Of course, these particles are different in their microscopic details and therefore quantum-mechanically distinguishable. Therefore, we cannot deduce from quantum statistical mechanics the factorial factor, which could be derived if they were quantum-mechanically indistinguishable. However, since we usually regard the particles as "identical" in experiments, we need the factorial to consistently explain experimental results (see, e.g., Refs. [18, 19]). Thus, the classical limit of quantum mechanics fails to describe thermodynamics of a colloidal system.

As we have discussed in Sect. 5.2.1, whether we regard particles as identical or different is at our disposal and each decision leads to its own thermodynamics. Each thermodynamics is self-consistent within its level of description as in the case of "identical" colloidal particles. Adherence to distinguishability in quantum mechanics spoils this beautiful universality of thermodynamics.

5.2.4.2 Logical Flaw

As we have discussed in the previous section, the reason why we identify the statistical-mechanical entropy as the thermodynamic entropy is that they have in common the thermodynamic relation (5.18). This reasoning is no less significant in quantum statistical mechanics than in classical statistical mechanics. Therefore, the entropy $S^{\mathrm{q\text{-}stat}}$ inevitably has the ambiguity function as

$$S(E, V, N) = S^{\mathrm{q}\text{-}stat}(E, V, N) + k_{\mathrm{B}} f^{\mathrm{q}}(N). \tag{5.27}$$

The conventional assumption of identifying $S^{\mathrm{q\text{-}stat}}$ with S is equivalent to assuming $f^{\mathrm{q}}(N) = 0$; however, this assumption has no physical basis. From a viewpoint of operational physics, we should go through the procedure to determine $f^{\mathrm{q}}(N)$, as we have done for classical statistical mechanics by requiring extensivity. Nontrivially, this procedure leads to a trivial consequence: $f^{\mathrm{q}}(N) = N f^{\mathrm{q}}(1)$, as we can explicitly confirm in the case of an ideal gas. In this sense, "the Gibbs paradox is no different in quantum mechanics, it is only less manifest" [6, p. 311].

5.3 Gibbs Paradox Revisited from the Fluctuation Theorem

As we see in the previous section, the classic problem of the Gibbs paradox is classified into the three aspects and they have completely been settled in the thermodynamic limit. In view of growing interests in thermodynamics of small systems, it would be worthwhile revisiting these issues in this context.

The resolution of GP-I to explain the counterintuitive behavior of the thermodynamic entropy upon gas mixing does not depend on the system size. Hence, it can be applied to small thermodynamic systems. The resolution of GP-II has already been given for the systems with a finite N and is therefore valid for small thermodynamic systems. However, the resolution of GP-III breaks down in small thermodynamic systems, because the assumption of extensivity is no longer valid; GP-III is open in small thermodynamic systems. As to this point, Jaynes [7, p. 17] expresses his concern by stating,

> The Pauli correction was an important step in the direction of getting 'the bulk of things' right pragmatically; but it ignores the small deviations from extensivity that are essential for treatment of some effects; and in any event it is not a fundamental theoretical principle. A truly general and quantitatively accurate definition of entropy must appeal to a deeper principle which is hardly recognized in the current literature [...].

Here, we show that in small thermodynamic systems the fluctuation theorem with absolute irreversibility takes the place of extensivity. To be specific, we show that the validity of the fluctuation theorem with absolute irreversibility is equivalent to a specific choice of the ambiguity function $f(N)$ as

$$[\text{thermodynamics with } \langle e^{-\beta(W-\Delta F)}\rangle = 1-\lambda] \Leftrightarrow f(N) = Nf(1) - \ln N!, \tag{5.28}$$

for classical statistical mechanics based on the specific phase. We note that the logical structure is completely parallel to the Pauli-Jaynes resolution (5.26).

To derive this result, we assume that inter-particle interactions do not break additivity of the thermodynamic entropy. By additivity, we mean that the thermodynamic entropy of a system consisting of independent subsystems is equal to the sum of the thermodynamic entropy of each subsystem. Note that extensivity implies additivity but not vice versa.

5.3.1 *Proof of the Main Claim*

First of all, we assume the fluctuation theorem with absolute irreversibility

$$\langle e^{-\beta(W-\Delta F)}\rangle = 1-\lambda \tag{5.29}$$

to determine the ambiguity function. To this aim, we compare the thermodynamic entropy productions in two gas-mixing processes as illustrated in Fig. 5.3. Suppose that gases are initially in thermal equilibria with inverse temperature β and the numbers of particles in two boxes are given by M and N, respectively. We denote the volumes of the boxes by V_{L} and V_{R}. Moreover, we assume that interactions between different particles are identical to those between identical particles. In the following, we calculate the difference of the entropy productions in these gas mixings in two ways, i.e., from the fluctuation theorem and from statistical mechanics, and by comparing these results we obtain a relation between the ambiguity function and the degree of absolute irreversibility. Therefore, by evaluating the degree of absolute irreversibility, we determine the ambiguity function.

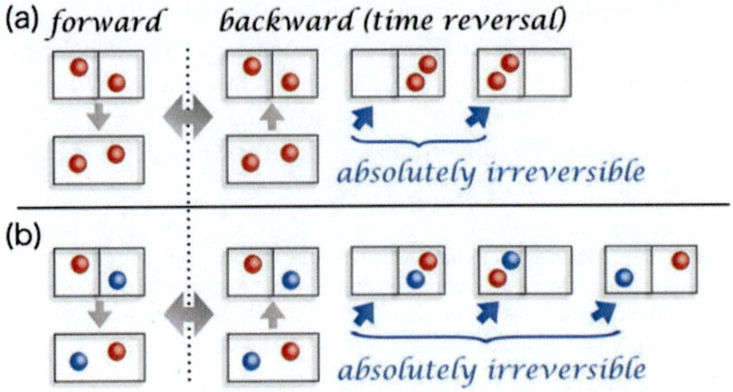

Fig. 5.3 Two gas mixings and their time reversal. **a** Identical-gas mixing and its time reversal. The time reversal of gas mixing is wall insertion for the global equilibrium state over the entire box. The particle-number distribution after the wall insertion can be different from the original state of the forward process as in the rightmost two cases. These cases contribute to absolute irreversibility. **b** Different gas mixing and its time reversal. The rightmost event is absent in **a**. Hence, different-gas mixing has a larger amount of absolute irreversibility. Reprinted figure with permission from Ref. [1]. Copyright 2017 by the American Physical Society

Let us calculate the entropy productions from the fluctuation theorem. For the identical-gas mixing, the fluctuation theorem reads

$$\langle e^{-\beta(W_{\rm id}-\Delta F_{\rm id})}\rangle_{\rm id} = 1-\lambda_{\rm id}. \tag{5.30}$$

Since $W_{\rm id}$ identically vanishes in this classical gas mixing, we obtain

$$\Delta F_{\rm id} = k_{\rm B}T\ln(1-\lambda_{\rm id}). \tag{5.31}$$

From the thermodynamic relation

$$\Delta F_{\rm id} = \Delta E_{\rm id} - T\Delta S_{\rm id}, \tag{5.32}$$

the thermodynamic entropy production is determined as

$$\Delta S_{\rm id} = \frac{1}{T}\Delta E_{\rm id} - k_{\rm B}\ln(1-\lambda_{\rm id}). \tag{5.33}$$

In the same way, we obtain the entropy production in the different-gas mixing as

$$\Delta S_{\rm dif} = \frac{1}{T}\Delta E_{\rm dif} - k_{\rm B}\ln(1-\lambda_{\rm dif}). \tag{5.34}$$

By the assumption of the identical interactions, the changes in the internal energy in the two cases are the same:

$$\Delta E_{\mathrm{id}} = \Delta E_{\mathrm{dif}}. \tag{5.35}$$

Therefore, the difference between the thermodynamic entropy productions is given by

$$\Delta S_{\mathrm{id}} - \Delta S_{\mathrm{dif}} = k_{\mathrm{B}} \ln \frac{1-\lambda_{\mathrm{dif}}}{1-\lambda_{\mathrm{id}}}. \tag{5.36}$$

Now, let us evaluate the same quantity from classical statistical mechanics based on the specific phase. For identical-gas mixing, the thermodynamic entropy production is given by

$$\Delta S_{\mathrm{id}} = \Delta S_{\mathrm{id}}^{\mathrm{stat}} + \Delta f_{\mathrm{id}}. \tag{5.37}$$

In the initial state, the ambiguity function is $f_{\mathrm{id}}^{\mathrm{ini}} = f(M) + f(N)$ by the assumption of additivity. Since the final state has the ambiguity function $f(M+N)$, we obtain

$$\Delta f_{\mathrm{id}} = f(M+N) - f(M) - f(N). \tag{5.38}$$

On the other hand, for different-gas mixing, we have

$$\Delta S_{\mathrm{dif}} = \Delta S_{\mathrm{dif}}^{\mathrm{stat}} + \Delta f_{\mathrm{dif}}. \tag{5.39}$$

Since we can quasistatically connect the initial and the final states by invoking the process with semi-permeable walls, the ambiguity function should remain unchanged: $\Delta f_{\mathrm{dif}} = 0$. Since the Hamiltonians give the same values because of the assumption of identical interactions, the integrals of the Boltzmann factor are also the same in the two cases. In classical statistical mechanics based on the specific phase, they constitute the partition functions since we give no correction about particle numbers. Therefore, the statistical mechanical entropies calculated from these identical partition functions are the same:

$$\Delta S_{\mathrm{id}}^{\mathrm{stat}} = \Delta S_{\mathrm{dif}}^{\mathrm{stat}}. \tag{5.40}$$

Hence, the difference of the thermodynamic entropy productions can be evaluated only in terms of the ambiguity function as

$$\Delta S_{\mathrm{id}} - \Delta S_{\mathrm{dif}} = k_{\mathrm{B}}[f(M+N) - f(M) - f(N)]. \tag{5.41}$$

Combining Eqs. (5.36) and (5.41), we acquire

$$f(M+N) - f(M) - f(N) = -\ln \frac{1-\lambda_{\mathrm{id}}}{1-\lambda_{\mathrm{dif}}}. \tag{5.42}$$

Therefore, the ambiguity in the entropy $f(N)$ is to be removed by the degrees of absolute irreversibility in the two gas-mixing processes.

To evaluate the degrees of absolute irreversibility, we consider the time-reversed process of gas mixing, i.e., wall insertion. Since the time-reversed process starts from the thermal equilibrium over the entire vessel, the particle-number distribution after the wall insertion fluctuates. The events in which the state does not return to the initial state is absolutely irreversible and hence contribute to λ. Conversely, $1 - \lambda$ is the probability that the state returns to the original one. Unfortunately, this probability cannot directly be evaluated because of the presence of interactions between particles. However, we can evaluate the ratio of the probabilities in the two cases. In identical-gas mixing, only the particle number should be restored (see Fig. 5.3a), whilst, in different-gas mixing, the species of the particle should be restored in addition. Therefore, the number of the restoring events in identical-gas mixing is larger than that in different-gas mixing by the factor of combinatorics to choose M out of $M + N$ particles. Since these events occur with the same probability due to the assumption of the same interactions, the ratio of the restoring probabilities is given by the binomial coefficient as

$$\frac{1 - \lambda_{\mathrm{id}}}{1 - \lambda_{\mathrm{dif}}} = \binom{M+N}{M}. \tag{5.43}$$

Therefore, we obtain the following functional equation:

$$f(M+N) - f(M) - f(N) = -\ln\binom{M+N}{M}. \tag{5.44}$$

Finally, let us solve this equation. By setting $M = 1$, we obtain

$$f(N+1) - f(N) = f(1) - \ln(N+1). \tag{5.45}$$

Summing up this equation, we acquire

$$\sum_{n=1}^{N-1}[f(n+1) - f(n)] = (N-1)f(1) - \sum_{n=1}^{N-1}\ln(n+1), \tag{5.46}$$

$$f(N) - f(1) = (N-1)f(1) - \ln N!. \tag{5.47}$$

Hence, we conclude

$$f(N) = Nf(1) - \ln N!. \tag{5.48}$$

Conversely, we now derive the fluctuation theorem with absolute irreversibility from the ambiguity function of the form (5.48). Let $Z(T, V, N)$ denote the partition function of the gas with temperature T, volume V, and particle number N calculated on the basis of classical statistical mechanics based on the specific phase. Then, the thermodynamic free energy is evaluated as

$$F(T, V, N) = -\beta^{-1}\ln Z(T, V, N) - \beta^{-1}f(N). \tag{5.49}$$

For identical-gas mixing, the change of the thermodynamic free energy is evaluated as

$$\Delta F_{\rm id} = F(T, V_{\rm L} + V_{\rm R}, M + N) - F(T, V_{\rm L}, M) - F(T, V_{\rm R}, N) \quad (5.50)$$

$$= -\beta^{-1} \ln \frac{Z(T, V_{\rm L} + V_{\rm R}, M + N)}{Z(T, V_{\rm L}, M) Z(T, V_{\rm R}, N)} + \beta^{-1} \ln \binom{M+N}{N}. \quad (5.51)$$

Therefore, we obtain

$$\langle e^{-\beta(W_{\rm id} - \Delta F_{\rm id})} \rangle = e^{\beta \Delta F_{\rm id}} \quad (5.52)$$

$$= \binom{M+N}{M} \frac{Z(T, V_{\rm L}, M) Z(T, V_{\rm R}, N)}{Z(T, V_{\rm L} + V_{\rm R}, M + N)}, \quad (5.53)$$

which is nothing but the probability to restore the original state by wall insertion; this value is equal to $1 - \lambda_{\rm id}$ by definition. Thus, the fluctuation theorem with absolute irreversibility is confirmed for identical-gas mixing. In a similar manner, it can be verified for different-gas mixing.

Hence, the equivalence relation (5.28) has been shown.

5.3.2 *Discussions*

We here give some related discussions.

5.3.2.1 Significance of Our Work

We have shown the equivalence between the requirement of the fluctuation theorem with absolute irreversibility and the specific choice of the ambiguity function with the factorial factor. Therefore, from a mathematical point of view, requiring either of them makes no difference. However, from a physical point of view, we believe that the requirement of the fluctuation theorem is more natural. Thermodynamics is a theory operationally defined on physical ground. The fluctuation theorem is described by operationally accessible quantities as the work distribution and the degree of absolute irreversibility. In this sense, the requirement of the fluctuation theorem would be more suitable as a foundation of operational thermodynamics.

5.3.2.2 Thermodynamic Criterion of Identity

As we see in the context of GP-I, thermodynamics depends on whether we are able and willing to distinguish particles. Therefore, to deal with the thermodynamic entropy, we have to introduce a criterion of whether we regard particles as identical or different. We do so by invoking different values of absolute irreversibility. Absolute irreversibility is evaluated as the probability that the state does not return to the original state by the time-reversed process. The thermodynamic criterion of identity of particles are given by our decision about whether we regard the returned state as the same as or different from the original state. In this regard, we are in line with the conventional wisdom to resolve the Gibbs paradox.

5.3.2.3 Relation Between Extensivity and the Fluctuation Theorem

In the thermodynamic limit, the requirement of extensivity leads to the desired form of the ambiguity function. Meanwhile, in small thermodynamic systems, the fluctuation theorem with absolute irreversibility takes the place of extensivity. Therefore, the requirement of the fluctuation theorem should reduce to the requirement of extensivity in the thermodynamic limit.

For simplicity, let us consider mixing of ideal identical gases with $M = N$ and $V_L = V_R =: V$. In this case, the degree of absolute irreversibility can explicitly evaluated as

$$\lambda_{\mathrm{id}} = 1 - \binom{2N}{N}\left(\frac{1}{2}\right)^{2N}. \tag{5.54}$$

Then, the thermodynamic entropy production

$$\Delta S_{\mathrm{id}} = -k_B \ln(1 - \lambda_{\mathrm{id}}) \simeq \frac{1}{2}\ln(\pi N) \tag{5.55}$$

is sub-extensive. In the thermodynamic limit, such a sub-extensive quantity is ignored. Thus, the requirement of the fluctuation theorem with absolute irreversibility leads to no entropy production upon identical-gas mixing as

$$S(T, V, N|T, V, N) = S(T, 2V, 2N). \tag{5.56}$$

When we assume additivity, the entropy of the initial state is the sum of the entropy of each part: $S(T, V, N|T, V, N) = 2S(T, V, N)$. Therefore, we obtain

$$S(T, 2V, 2N) = 2S(T, V, N), \tag{5.57}$$

which is extensivity (5.23) for $q = 2$. Thus, the fluctuation theorem with absolute irreversibility implies extensivity in the thermodynamic limit.

5.3.2.4 Effects of Interactions

We deal with nonvanishing interactions as long as they do not break additivity. In the presence of interactions, such quantities as the energy change ΔE and entropy production ΔS cannot explicitly evaluated. However, by comparing the two cases of gas mixing, we cancel out these nontrivial effects of interactions and as a consequence derive the relation between the ambiguity function and the degrees of absolute irreversibility. Thus, our strategy to compare the two mixing processes plays a vital role in removing the ambiguity in the presence of interactions.

In the Pauli-Jaynes resolution, one has to invoke a concrete formula of the entropy calculated by statistical mechanics. Therefore, the extension of this method to a general interacting gas seems intractable. In this sense, our method is superior to the conventional method even in the thermodynamic limit.

5.3.2.5 Why Not Start from an Open System

One may think that we can determine the particle-number dependence of the entropy if we start from an open system from the beginning. We could extend the Clausius equality to an open system as

$$\Delta S = \int_{\text{quasistatic}} \frac{\delta Q + \mu dN}{T}. \tag{5.58}$$

However, to use this formula as the definition of the entropy, we have to define the chemical potential μ beforehand. Unfortunately, the chemical potential is defined as the particle-number dependence of the thermodynamic entropy as

$$\mu = \frac{\partial S}{\partial N}. \tag{5.59}$$

Therefore, we end up with a circular argument when we start from an open system.

5.3.2.6 Isothermal Versus Isolated Systems

We discuss the Gibbs paradox in isothermal systems instead of isolated systems. One reason is that Gibbs originally considered the isothermal situation at a constant temperature. Another important reason is that microcanonical ensemble in statistical mechanics only gives an approximate description of thermodynamics for small isolated systems. On the other hand, when we consider isothermal small systems attached to infinitely large thermal environments, statistical-mechanical description should have perfect accuracy since the microcanonical description for the infinite combined system is accurate. Therefore, the consistency problem between thermodynamics and statistical mechanics is more well defined in isothermal systems than in isolated systems.

References

1. Murashita Y, Ueda M (2017) Phys Rev Lett 118:060601
2. Murashita Y (2015) Absolute irreversibility in information thermodynamics. Master's thesis, The University of Tokyo
3. Gibbs JW (1875-78) On the equilibrium of heterogeneous substances. Connecticut Academy of Science
4. Gibbs JW (1902) Elementary principles in statistical mechanics. Yale University Press
5. Ehrenfest P, Trkal V (1920) Ann d Phys 65:609
6. van Kampen NG (1984) The Gibbs paradox. Pergamon
7. Jaynes ET (1992) The Gibbs paradox. Kluwer Academic Publishers
8. Sackur O (1911) Ann d Phys 36:958
9. Sackur O (1913) Ann d Phys 40:67
10. Tetrode H (1912) Ann d Phys 38:434
11. Pauli W (1973) Thermodynamics and the kinetic theory of gases. MIT Press
12. Feynman RP (1972) Statistical mechanics: a set of lectures. Westview Press
13. Zubarev D (1974) Nonequilibrium statistical thermodynamics. Plenum Publishing Corporation
14. Landau LD, Lifshitz EM (1975) Statistical physics. Butterworth-Heinemann
15. Toda M, Kubo R, Saito N (1978) Statistical physics I: equilibrium statistical mechanics. Springer
16. Callen HB (1985) Thermodynamics and an introduction to thermostatistics. Wiley
17. Kondepudi D, Prigogine I (1998) Modern thermodynamics: from heat engines to dissipative structures. Wiley
18. Frenkel D (2014) Mol Phys 112:2325
19. Cates ME, Manoharan VN (2015) Soft Matter 11:6538

Chapter 6
Loschmidt Paradox and the Fluctuation Theorems

In this chapter, we show a close relationship between the Loschmidt irreversibility paradox and the fluctuation theorem with absolute irreversibility. First of all, we review some historical discussions by Boltzmann and Loschmidt. Then, we see that fractality plays a crucial role in resolving the Loschmidt paradox in dissipative reversible chaotic dynamics. Next, we extend this fractal scenario of emergent irreversibility to conservative (i.e., nondissipative) reversible chaotic dynamics. We use the Rényi divergence to evaluate fractality in phase space of a dynamical billiard, and find that fractality emerges in an intermediate time regime despite the Liouville theorem. Finally, we reformulate our discussion in view of the fluctuation theorem. As a result, we give a lower bound for the information-theoretic irreversibility, which is determined by the degree of absolute irreversibility. Most of the contents of this paper is based on the published article [1].

6.1 Historical Discussions

Boltzmann derived the time-evolution equation of the probability density in phase space and showed on the basis of it that the H-function, which is the negative of the entropy, monotonically decreases in time [2]. However, Loschmidt pointed out that, if one has a trajectory, one should have its time-reversed trajectory under reversible dynamics [3]. Boltzmann reformulated what Loschmidt implied in thermodynamic terms in a transparent manner, and argued that events with a negative entropy production occur much less frequently than those with a positive entropy production despite the one-to-one correspondence between them [4].

Y. Murashita, *Fluctuation Theorems under Divergent Entropy Production and their Applications for Fundamental Problems in Statistical Physics*, Springer Theses, https://doi.org/10.1007/978-981-16-8638-2_6

6.1.1 Boltzmann's Equation

In Ref. [2], Boltzmann considered the time evolution of a dilute monoatomic gas. We here consider the probability distribution in phase space $f(p, q)$, although Boltzmann originally considered the time evolution of the probability distribution of the kinetic energy [2]. Then, the time evolution of $f(p, q)$ can be shown to be governed by Boltzmann's equation. One of the important consequences of Boltzmann's equation is the H-theorem. The H-function is defined as

$$H[f] = \int dpdq\ f(p, q) \ln f(p, q). \tag{6.1}$$

Using Boltzmann's equation, one can show that $H[f]$ is non-increasing and stays constant if and only if $f(p, q)$ is a canonical distribution.

If $f(p, q)$ is given by a canonical distribution and the gas is so dilute that the interaction energy between gas molecules is negligible, then $H[f]$ coincides with the negative of the thermodynamic entropy per particle in units of k_{B} up to an arbitrary additive constant. Therefore, H can be regarded as a nonequilibrium extension of the equilibrium thermodynamic entropy, or at least a function that characterizes the approach to equilibrium. In fact, Boltzmann concluded [2, p. 306],

> Es ist somit strenge bewiesen, dass, wie immer die Vertheilung der lebendigen Kraft zu Anfang der Zeit gewesen sein mag, sie sich nach Verlauf einer sehr langen Zeit immer nothwendig der von Maxwell gefundenen nähern muss.

or

> Consequently, one has rigorously proved that, whatever the distribution of the kinetic energy at the initial time may have been, the state must always necessarily approach the one found by Maxwell after a very long time. (translated into English by the author)

However, there is an essential hidden assumption in his derivation of Boltzmann's equation. To calculate the evolution of the probability distribution due to collisions of gas molecules, Boltzmann assumed that the frequency of the collision depends only on the kinetic energies of the colliding particles, which implies that the memory of past collisions is completely lost. Since this assumption, which is now called the molecular chaos hypothesis, is crucial, Boltzmann's derivation of irreversibility from reversible dynamics leaves room for closer scrutiny.

6.1.2 Loschmidt Paradox

In his article [3, pp. 137–139], Loschmidt considered the thermal equilibrium state under the gravity. Suppose that one has a vessel filled with gas molecules. As an initial condition, we assume that one of the gas molecules rests on the ceiling and the others rest on the bottom. Due to the gravity, the particle on the ceiling falls and collides with other particles. After many collisions between particles, the velocity distribution is expected to finally equilibrate to become a thermal one.

However, Loschmidt argued that this observation has a limitation. Let us assume that one indeed gets an equilibrated state after time τ from the above-described initial state. Suppose that one then changes the directions of velocities of all particles. Then, the obtained state is an equilibrium state with the thermal velocity distribution. However, if one lets the system evolve over time τ, the state ends up with an extremely nonthermal state; one particle absorbs all the energy of other particles and reaches the ceiling of the vessel.

Boltzmann reformulated this discussion in a more transparent manner [4]. Suppose that one has an initial condition that leads to a positive entropy production. Then, one can construct an initial condition that leads to a negative entropy production by using the method that Loschmidt proposed; one reverses all velocities of the final state of the original event with a positive entropy production. Therefore, any attempt to prove the second law of thermodynamics should involve considerations on the initial condition. In fact, Boltzmann stated [4, Sect. 2],

> Man kann also nicht beweisen, daß, wie immer die Positionen und Geschwindig- keiten der Kugeln zu Anfang gewesen sein mögen, nach Verlauf einer sehr langen Zeit die Verteilung immer gleichförmig sein muß, sondern bloß, daß unendlich vielmal mehr Anfangszustände nach Verlauf einer bestimmten längeren Zeit zu einer gleichförmigen, als zu einer ungleichförmigen Zustandsverteilung führen und daß auch im letzteren Falle nach Verlauf einer noch längeren Zeit die Zustandsverteilung wieder gleichförmig wird.

or

> One cannot therefore prove that, whatever the initial positions and velocities of the balls may have been, after a very long time the distribution must always be uniform; one can merely prove that initial states that after a certain long time lead to a uniform state are infinitely much more than those that lead to a nonuniform state, and that in the latter case after an even longer time the state again becomes uniform. (translated into English by the author)

6.2 Resolution by Fractality

After more than a century, Boltzmann's observation that positive-entropy-producing events are infinitely more frequent than negative-entropy-producing ones was confirmed. Holian et al. [5] conducted numerical simulations using the Nosé-Hoover equation, which is a reversible but dissipative equation of motion. They found that the initial states that lead to a negative entropy production reside on a zero-volume fractal attractor and therefore improbable. This observation can be quantitatively formulated in light of the steady-state fluctuation theorem [6–9], which is derived from the invariant measure on the fractal. Thus, fractality in phase space plays a key role in deriving emergent irreversibility from dissipative reversible dynamics.

6.2.1 Fractality from Reversible Dynamics

Holian et al. [5] considered the origin of irreversibility in nonequilibrium molecular dynamics. They invoked the Nosé-Hoover equation [10–12] to simulate heat conduction. The Nosé-Hoover equation is a deterministic set of equations on phase space (p, q) written as

$$\dot{p}_t = F(q_t) - \gamma_t p_t, \tag{6.2}$$

$$m\dot{q}_t = p_t, \tag{6.3}$$

where m and $F(q_t)$ are the mass and the conservative force, respectively. The friction coefficient γ_t represents the degree of freedom of the heat bath and its time evolution is given by

$$\dot{\gamma}_t = \left(\frac{\sum m\dot{q}_t^2}{X k_{\mathrm{B}} T} - 1 \right) / \tau^2, \tag{6.4}$$

where X is the number of degrees of freedom of the system and τ is the relaxation time. Thus, the degree of freedom of the heat bath is driven by the difference between the kinetic energy and its thermal value with temperature T.

The Nosé-Hoover equations of motion are time-reversal symmetric; if one has a solution (p_t, q_t, γ_t) from time $t = 0$ to τ, one has the corresponding time-reversed solution $(-p_{\tau-t}, q_{\tau-t}, -\gamma_{\tau-t})$ as one can explicitly verify from Eqs. (6.2)–(6.4). Thus, the time-reversed trajectory obeys the identical equations of motion. This behavior is completely different from phenomenological dissipative equations of motion, such as the diffusion equation and chaotic maps, which connote irreversibility with respect to time from the beginning.

As an illustrative nonequilibrium system, one can consider a bulk system with two boundary regions [5]. In the bulk, particles evolve according to the Newton equations of motion, while, on the two boundaries, particles evolve in accordance with the Nosé-Hoover equations. The temperatures of the two boundaries are set to be different from each other; the Nosé-Hoover equations with different temperatures apply to the boundaries. Then, from numerical simulations, one can see that heat flows from the high-temperature region to the low-temperature one regardless of the initial condition. Therefore, the Nosé-Hoover equations show macroscopic dissipative irreversible behavior despite their original time reversibility.

In view of phase space, this irreversible behavior corresponds to contraction in phase space. The time evolution leads states in phase space to collapse onto a subspace with zero volume, namely, a fractal attractor. Then, if one wants the final state whose entropy decreases in time by the velocity reversal, one should sample it from the zero-volume fractal. Therefore, the probability to sample such second-law-violating events is zero. By choosing the state near the fractal, one can violate the second law for a certain time interval corresponding to the closeness to the fractal attractor. Nevertheless, one can permanently violate the second law only if one samples the initial state that is exactly on the fractal attractor with zero volume. Therefore, it can be said that [5, p. 13]

> unstable states going backward in time are never observed, *not* because they violate the equations of motion, but rather because the probability of observing them is precisely zero.

6.2.2 *Steady-State Fluctuation Theorem*

The fractal resolution of the Loschmidt irreversibility paradox given in the previous section can be quantitatively formulated in terms of fluctuation theorems. For reversible dissipative chaotic dynamics, the following steady-state fluctuation theorem holds [6–9]:

$$\frac{P(-\sigma)}{P(\sigma)} \simeq e^{-\sigma}, \tag{6.5}$$

where σ represents the entropy production, and $P(\cdot)$ is its probability distribution function. Equation (6.5) implies that the probability for a negative entropy production is exponentially suppressed compared with that for the positive entropy production. Since the entropy production is extensive in time in a steady state, the negative-entropy-producing probability exponentially decreases in time. In particular, it vanishes in the long-time limit. Thus, the steady-state fluctuation theorem indicates emergent irreversible behavior from reversible chaos. It is noteworthy that the steady-state fluctuation theorem is derived from the invariant measure with the fractal support. Fractality again plays a crucial role.

6.3 Transient Fractality in a Chaotic Hamiltonian System

As we have seen in the previous sections, fractality is the origin of irreversibility in dissipative reversible chaotic dynamics. We would like to address the question of whether this fractal scenario of irreversibility applies to the conservative reversible chaotic dynamics, in particular, to the chaotic Hamiltonian dynamics. One may think that the answer should be no since the Liouville theorem, or the conservation law of phase-space volume, seems incompatible with the zero-volume fractal attractor. Nevertheless, we have found that a fractal structure transiently emerges in an intermediate length scale in the Hamiltonian dynamics.

First of all, we introduce our method of evaluating fractality in terms of Rényi divergence. Then, we consider the Bunimovich billiard as a paradigmatic chaotic Hamiltonian system, and show that a foliate structure emerges by the time evolution. Next, to evaluate the fractality of the generated structure by the Rényi divergence, we introduce the time-reversal test by an imperfect Loschmidt demon. Finally, we show that from the obtained values of the Rényi divergence we can see the emergence of a fractal scaling law in an intermediate time and length scales.

6.3.1 Rényi-0 Divergence and Fractal Dimension

To evaluate fractality of a structure ρ in the embedding space with the dimension d_{E}, we assess its vulnerability against perturbation. To be precise, we consider the convolution of ρ and the d_{E}-dimensional isotropic Gaussian with variance l, namely,

$$\mathscr{C}_l[\rho](\gamma) := \int \rho(\gamma') \left(\frac{1}{\sqrt{2\pi l^2}} \right)^{d_{\mathrm{E}}} \exp \left[-\frac{(\gamma - \gamma')^2}{2l^2} \right] d\gamma, \tag{6.6}$$

where γ and γ' are d_{E}-dimensional vectors in the embedding space. Then, we evaluate the difference between the original state and the perturbed state by the Rényi-0 divergence [13]:

$$D_0(\rho||\mathscr{C}_l[\rho]) := -\ln \int_{\rho(\gamma)>0} \mathscr{C}_l[\rho](\gamma) d\gamma. \tag{6.7}$$

As shown in Appendix 1, when ρ has a fractal structure with the fractal dimension d_{F}, the Rényi-0 divergence satisfies

$$D_0(\rho||\mathscr{C}_l[\rho]) \simeq d_{\mathrm{C}} \ln \frac{l}{l_0}, \tag{6.8}$$

where d_{C} is the fractal codimension defined by

$$d_{\mathrm{C}} := d_{\mathrm{E}} - d_{\mathrm{F}}, \tag{6.9}$$

and l_0 represents the length scale of the finest structure of ρ. Therefore, the linear dependence of D_0 on $\ln l$ is a signature of fractality. Moreover, from the l-dependence of D_0, we can extract the fractal dimension of the structure. Although the ordinary discussions consider fractality only in the limit of $l \to 0$, we also consider fractality in an intermediate length scale $l(\neq 0)$.

6.3.2 Bunimovich Billard as a Paradigmatic Example

As a paradigmatic chaotic Hamiltonian system, we consider the Bunimovich billiard [14]. In this system, a particle ballistically moves and elastically and specularly bounces on the wall like the stadium as shown in Fig. 6.1a. Due to the conservation of the energy, phase space of the Bunimovich billiard can be taken to be three-dimensional ($d_{\mathrm{E}} = 3$). A typical choice of the coordinates consists of two positions (x and y) and an angle of the velocity (θ).

For illustration, we prepare the initial state ρ_0 as a uniform distribution over a cube in phase space as shown in Fig. 6.1b. Then, we let it evolve over time T to obtain

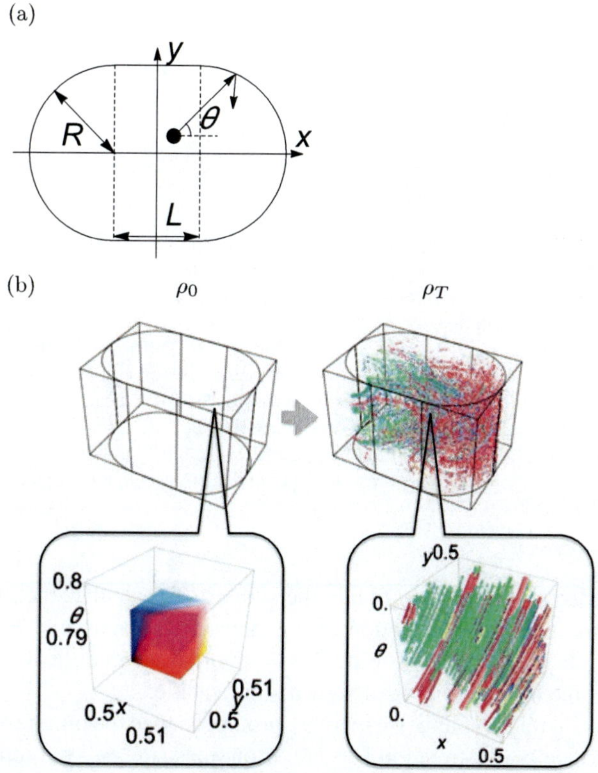

Fig. 6.1 Bunimovich billiard and time evolution. **a** Geometry of the Bunimovich stadium. A dynamical billiard ballistically moves and elastically and specularly bounces on the wall of the stadium. Due to the energy conservation, phase space can be represented by two positions x and y, and the angle of velocity θ. **b** Foliate structure in phase space. As the initial state ρ_0, a uniform distribution over a cube is prepared. Then, we let it evolve over time T, and obtain the final distribution ρ_T. As we can see in the amplified view, it has a foliate structure, which is characteristic of fractals. The geometrical parameters are set as $R = 1$ and $L = 1$. The evolution time is $T = 30$. The initial state ρ_0 is a uniform distribution over the cube with the side length $w = 0.01$ and the least-valued vertex $(x_0, y_0, \theta_0) = (0.5, 0.5, \pi/4)$. Reprinted figure from Ref. [1] under the Creative Commons Attribution 4.0 International license

the final state ρ_T. Remarkably, the obtained state has a foliate structure, which is a unique feature of fractals.

6.3.3 *Time Reversal Test: Imperfect Loschmidt Demon*

As we can see in Fig. 6.1b, the final state ρ_T has an extremely complicated support. Since the support is exponentially elongated in time due to chaos, exponential trials

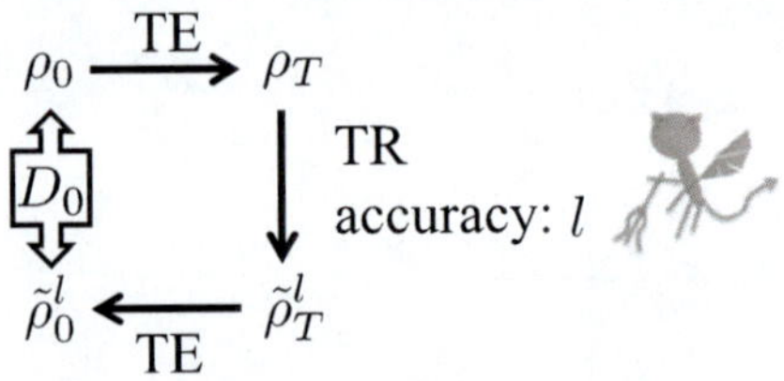

Fig. 6.2 Protocol of the time-reversal test. The initial state ρ_0 time-evolves (TE) into the final state ρ_T. Since it is difficult to directly evaluate the Rényi-0 divergence between ρ_T and $\mathscr{C}_l[\rho_T]$, we utilize the protocol of the time-reversal (TR) test. We prepare the velocity-reversed state $\tilde{\rho}_T^l = \mathscr{T}\mathscr{C}_l[\rho_T]$ with accuracy l, where $\mathscr{T}$ represents the reversal of the velocity. Then, we let it evolve during the same time interval T to obtain the state $\tilde{\rho}_0^l$, which is imperfectly pulled back to the initial time due to an imperfect accuracy upon time reversal. As a result, the Rényi-0 divergence can be evaluated at the initial time since $D_0(\rho_0||\mathscr{T}\tilde{\rho}_0^l) = D_0(\rho_T||\mathscr{T}\tilde{\rho}_T^l)$ (see Eq. (6.10)). Since ρ_0 can be chosen to have a simple support, the evaluation of D_0 at the initial time can be much simpler and, in fact, becomes numerically tractable. Thus, the fractality of ρ_T can be evaluated by the Loschmidt demon with no absolute perfection. Reprinted figure from Ref. [1] under the Creative Commons Attribution 4.0 International license

would be needed to numerically obtain the support. To calculate the Rényi-0 divergence, we should count the fraction of the states that remain in the support after the perturbation with strength l. Since the support itself is difficult to obtain, the calculation of the Rényi-0 divergence seems intractable.

To circumvent this problem, we invoke a protocol called the time-reversal test [15, 16] as schematically illustrated in Fig. 6.2. From the state ρ_T, we prepare the time-reversed state $\tilde{\rho}_T^l = \mathscr{T}\mathscr{C}_l[\rho_T]$ with accuracy l, where the operator $\mathscr{T}$ represents time reversal or velocity inversion. Then, we pull back the state to the initial time by the reversed time evolution to obtain $\tilde{\rho}_0^l$. Since

$$D_0(\rho_T||\mathscr{C}_l[\rho_T]) = D_0(\rho_T||\mathscr{T}\tilde{\rho}_T^l) = D_0(\rho_0||\mathscr{T}\tilde{\rho}_0^l), \tag{6.10}$$

the Rényi-0 divergence can be evaluated at the initial time. When we choose ρ_0 as such a simple state as the uniform distribution over a cube, the evaluation of $D_0(\rho_0||\mathscr{T}\tilde{\rho}_0^l)$ is tractable. Thus, the fractality of ρ_T can be evaluated by the time-reversal test or the protocol of the imperfect Loschmidt demon with accuracy l.

In Fig. 6.3, the time reversal test is conducted in the Bunimovich billiard, where we prepare the uniform distribution over a cube as the initial state. Although the difference between ρ_T and $\mathscr{T}\tilde{\rho}_T^l = \mathscr{C}_l[\rho_T]$ cannot be discerned, it can be manifested as the difference between the initial state ρ_0 and the state $\mathscr{T}\tilde{\rho}_0^l$, which is pulled back to the initial time with accuracy l. When T is short, these states are almost the same up to the velocity inversion provided that l is much smaller than the side of the cube w. As a result, the Rényi-0 divergence D_0 is almost zero. As time proceeds, the pulled-back state $\tilde{\rho}_0^l$ becomes more and more different from the initial state ρ_0, resulting in larger Rényi-0 divergence. Finally, $\tilde{\rho}_0^l$ becomes almost uniform over the entire phase space. Consequently, the value of the Rényi-0 divergence saturates to the value corresponding to the volume ratio of the entire phase space to the initial

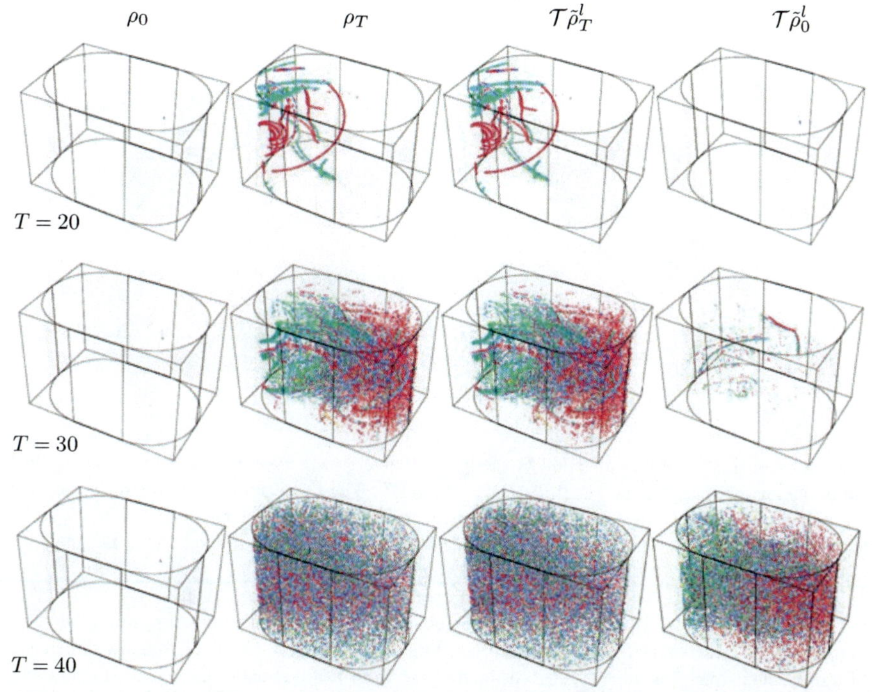

Fig. 6.3 Time reversal test in the Bunimovich billiard by the imperfect Loschmidt demon. When the evolution time T is short, the state $\tilde{\rho}_0^l$ pulled back to the initial time by the imperfect Loschmidt demon is almost identical with the true initial state ρ_0 up to the velocity inversion $\mathscr{T}$. However, as time proceeds, the structure of ρ_T becomes increasingly more complicated, resulting in greater difference between ρ_0 and $\mathscr{T}\tilde{\rho}_0^l$. This difference can be quantified by the Rényi-0 divergence $D_0(\rho_0||\mathscr{T}\tilde{\rho}_0^l)$. When T is large, the pullbacked state $\tilde{\rho}_0^l$ becomes almost uniform. Even if ρ_T and $\mathscr{T}\tilde{\rho}_T^l$ appear to be identical, they are actually very different for large T, as uncovered by the comparison between ρ_0 and $\mathscr{T}\rho_0^l$. This is why we should invoke the time reversal test to evaluate the Rényi-0 divergence $D_0(\rho_T||\mathscr{T}\tilde{\rho}_T^l) = D_0(\rho_T||\mathscr{C}_l[\rho_T])$. The geometrical parameters and the initial state are set as in the caption of Fig. 6.1. The accuracy of the imperfect Loschmidt demon is $l = e^{-15}$. Reprinted figure from Ref. [1] under the Creative Commons Attribution 4.0 International license

state, namely, $D_0 \simeq \ln[V_{\rm ps}/w^3]$, where $V_{\rm ps} = 2\pi(2RL + \pi R^2)$ is the volume of the entire phase space. Therefore, this saturation can be regarded as a finite-size effect.

6.3.4 Transient Fractality

We numerically obtain the Rényi-0 divergence D_0 while varying the length scale l and the time interval T as shown in Fig. 6.4a. Detailed explanations of numerics are given in the Appendix 3. For a fixed time T, we can see that D_0 is proportional to $\ln l$. According to Eq. (6.8), this proportionality indicates fractality. To see this

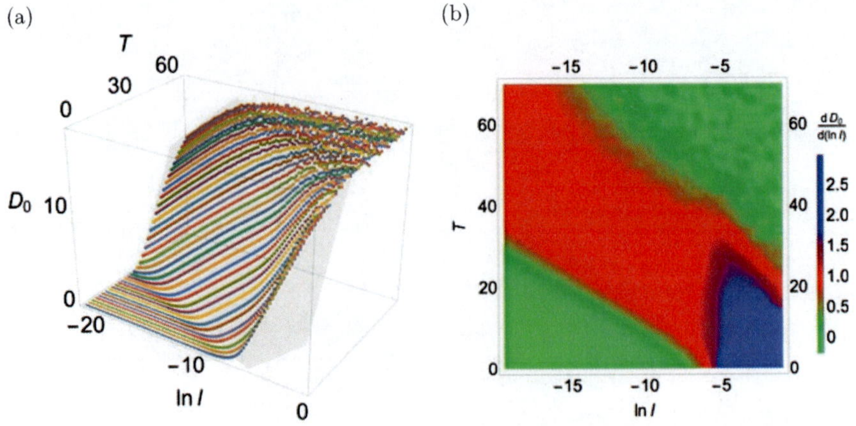

Fig. 6.4 Transient fractality unveiled by the Rényi-0 divergence. **a** Rényi-0 divergence D_0 against variations of the length scale l and the time interval T. The linear relation between D_0 and $\ln l$ for an intermediate length scale l and a fixed time T indicates that a fractal structure exists as indicated in Eq. (6.8). The grey plane is the theoretical estimate of the values of D_0, namely, Eq. (6.8) with $l_0 = we^{-\Lambda T}$, where Λ is the Lyapunov exponent of this system. **b** Derivative of the Rényi-0 divergence with respect to $\ln l$. For a fixed length scale l, there is a plateau in an intermediate time scale. This indicates that a fractal structure transiently emerges. Since the value of the plateau corresponds to the fractal codimension, we obtain $d_C = 1.0$. Therefore, the dimension of the transient fractal is $d_F = 2.0$. Reprinted figure from Ref. [1] under the Creative Commons Attribution 4.0 International license

property more clearly, we calculate the derivative of D_0 with respect to $\ln l$ as shown in Fig. 6.4b. For a fixed time T, the plateau corresponds to the linear region with fractality. When we fix the length scale l, we can see that fractality transiently emerge in an intermediate time scale. The value of the plateau corresponds to the fractal codimension, and therefore we can estimate $d_C = 1.0$. Hence, we conclude that the transient fractal with the dimension $d_F = 2.0$ dynamically emerges.

We note that the transient fractality found here is not forbidden by the Liouville theorem. Due to the Liouville theorem, the phase space volume is conserved and equal to that at the initial time. Meanwhile, a mathematical fractal with infinitely fine structures has zero volume. Therefore, these two concepts seem to contradict each other. However, we here consider fractality in a fixed length scale l. Therefore, the phase-space structure can be regarded as fractal as long as it has a finer structure than l, which does not necessarily have vanishing volume. Conversely, under a fixed l, a structure finer than l should be generated as time proceeds due to chaoticity. The finest structure generated by dynamics is approximately given by $l_0 \simeq we^{-\Lambda T}$, where w is the linear dimension of the initial cube, or more generally speaking the length scale of the initial state, and Λ is the unique positive Lyapunov exponent. We note that due to the time-reversal symmetry the Lyapunov spectrum can be written as $\{\Lambda, 0, -\Lambda\}$. When the observer with accuracy l becomes insensitive to the finest length scale l_0, the transient fractality begins to emerge. Therefore, the time scale for the onset of the transient fractality can be estimated to be $T_{\rm onset} \sim 1/\Lambda \cdot \ln[w/l]$.

Let us discuss why the transient fractality emerges with the dimension $d_{\rm F} = 2.0$. The Kaplan-Yorke method [17] enables us to estimate the fractal dimension from the Lyapunov spectrum (see Appendix 2 for details). From the present Lyapunov spectrum $\{\Lambda, 0, -\Lambda\}$ $(\Lambda > 0)$, this method gives the fractal dimension of $d_{\rm F} = 3$. This result is consistent with the observation that no fractal structure persists in the long-time limit. However, in the transient time, we numerically observe fractality with dimension $d_{\rm F} = 2$. To understand this behavior, we note that local structures in phase space are not affected by the positive Lyapunov exponent by the time when the phase-space structure looks with accuracy l as if it were self-crossed. Let $T_{\rm end}$ denote this time scale.[1] Then, before this time scale, the Lyapunov spectrum can effectively be regarded as $\{0, 0, -\Lambda\}$. From this Lyapunov spectrum and the Kaplan-Yorke method, we obtain the fractal dimension $d_{\rm F} = 2$. Thus, fractality with $d_{\rm F} = 2$ emerges during an intermediate time regime of $T_{\rm onset} \lesssim T \lesssim T_{\rm end}$.

We note that this transient fractal should be distinguished from the one called a fat fractal [18, 19]. The fat fractal has infinitely fine structure but its volume is nonzero.

By generalizing the above-described discussion, we can conjecture that, in a Hamiltonian system with the spatial dimension of d, a fractal structure with the dimension $d_{\rm F} = d$ transiently emerges. We describe further details in Appendix 2.

6.4 Reformulation in View of the Fluctuation Theorems

We reformulate the transient fractal scaling law in view of the fluctuation theorem by regarding the incomplete backward process as the reference process. By doing so, we derive the fluctuation theorem with absolute irreversibility and the corresponding second-law-like inequality, which gives a lower bound for information-theoretic irreversibility. The time evolution of this irreversibility is dominated by the Lyapunov exponent in a short time scale and then by the transient fractality in an intermediate time scale.

6.4.1 Imperfect Backward Process as the Reference Process

In the general formulation of the fluctuation theorem, we compare the path probability in the original process $\mathscr{P}[\Gamma]$ with the path probability in the reference process

[1] If we model the phase-space structure to be a self-avoiding walk of a two-dimensional ribbon with the width w and the length $we^{\Lambda T}$ in the meshes with the side l, the time when the entire phase space is filled by the ribbon is roughly estimated to be $1/\Lambda \cdot \ln[V_{\rm ps}/w^2 l]$, which gives the largest possible value of $T_{\rm end}$. Therefore, we have a rough estimate $T_{\rm end} \lesssim 1/\Lambda \cdot \ln[V_{\rm ps}/w^2 l]$. A more accurate estimate for $T_{\rm end}$ is yet to be found.

$\mathscr{P}^{\mathrm{r}}[\Gamma]$. Then, the detailed fluctuation theorem is written as

$$\frac{\mathscr{P}^{\mathrm{r}}[\Gamma]}{\mathscr{P}[\Gamma]} = e^{-\sigma[\Gamma]}, \tag{6.11}$$

where $\sigma[\Gamma]$ is the path-dependent entropy production. When we choose a proper reference process, the entropy production acquires a physical meaning. By extending this idea, we can utilize Eq. (6.11) as the definition of an empirical entropy production.

We here choose an imperfect backward process discussed in the previous sections as the reference process. Since the dynamics is deterministic, the path probability reduces to the probability of the initial state. Therefore, the empirical entropy production is given by

$$\sigma(\gamma_0) = -\ln \frac{\mathscr{T}\tilde{\rho}_0^l(\gamma_0)}{\rho_0(\gamma_0)}. \tag{6.12}$$

We note that the average $\langle\sigma\rangle$ is not the thermodynamic entropy production, but an information-theoretic measure called the Kullback-Leibler divergence or the Rényi-1 divergence defined by

$$D_1(\rho_0||\mathscr{T}\tilde{\rho}_0^l) = \int \rho_0(\gamma_0) \ln \frac{\rho_0(\gamma_0)}{\mathscr{T}\tilde{\rho}_0^l(\gamma_0)} d\gamma_0. \tag{6.13}$$

We also note that this quantity is invariant in time under the Hamiltonian dynamics as

$$D_1(\rho_0||\mathscr{T}\tilde{\rho}_0^l) = D_1(\rho_T||\mathscr{T}\tilde{\rho}_T^l). \tag{6.14}$$

Therefore, the average can be represented as

$$\langle\sigma\rangle = D_1(\rho_T||\mathscr{C}_l[\rho_T]), \tag{6.15}$$

which is an information-theoretic distance between the fractal state ρ_T and the state $\mathscr{C}_l[\rho_T]$ that the Loschmidt demon observes with nonzero accuracy l. Thus, the empirical entropy production σ gives an indicator of the information-theoretic distance.

6.4.2 *Fluctuation Theorem with Absolute Irreversibility*

Since the support of $\mathscr{T}\tilde{\rho}_0^l$ is not covered by that of ρ_0, we have situations where we divide by zero to form the probability ratio. Therefore, the fluctuation theorem involves the singular continuous absolute irreversibility. Consequently, the fluctuation theorem reads

$$\langle e^{-\sigma}\rangle = 1 - \lambda, \tag{6.16}$$

where the degree of absolute irreversibility is given by

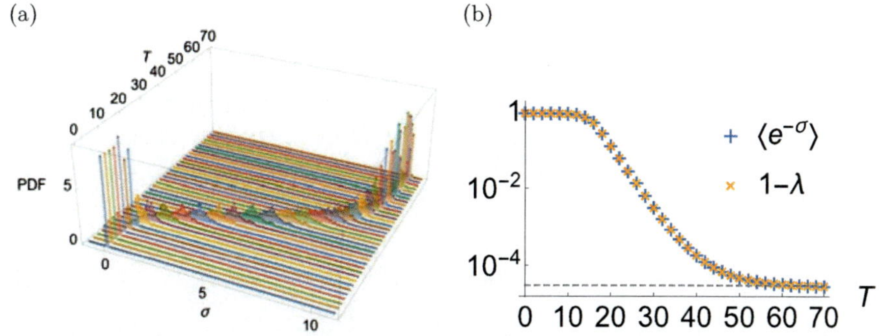

Fig. 6.5 Empirical entropy production and the fluctuation theorem with absolute irreversibility. **a** Probability distribution function (PDF) of the empirical entropy production σ for each time T. The values of the entropy production increase as time proceeds. Larger σ corresponds to stronger irreversibility in the time-reversal test. The values saturate due to a finite-size effect when the pullbacked state becomes almost uniform. **b** Verification of the fluctuation theorem with absolute irreversibility (6.16). The blue marks represent the averages $\langle e^{-\sigma} \rangle$ calculated from the PDF of σ. The yellow marks represent the values of the degree of absolute irreversibility obtained from the Rényi-0 divergence. The grey dashed line represents the level of the saturation. We set $w = 0.1$ and $l = e^{-10}$. Other parameters are the same as those in the caption of Fig. 6.1. Reprinted figure from Ref. [1] under the Creative Commons Attribution 4.0 International license

$$\lambda = \int_{\rho(\gamma_0)=0} \mathscr{T} \tilde{\rho}_0^l(\gamma_0) d\gamma_0. \tag{6.17}$$

Interestingly, this quantity has a direct relation to the Rényi-0 divergence as

$$1 - \lambda = e^{-D_0}. \tag{6.18}$$

By applying Jensen's inequality to the fluctuation theorem (6.16), we obtain a second-law-like inequality

$$\langle \sigma \rangle \geq -\ln(1 - \lambda). \tag{6.19}$$

In this case, this inequality is equivalent to the inequality known in the information theory (see, e.g., Ref. [13]):

$$D_1(\rho_T || \mathscr{C}_l[\rho_T]) \geq D_0(\rho_T || \mathscr{C}_l[\rho_T]). \tag{6.20}$$

From a numerical point of view, D_0 is more accessible than D_1.

To verify the fluctuation theorem numerically, we calculate the probability distribution function of the empirical entropy production σ as shown in Fig. 6.5a. See Appendix 3 for details. We can observe that the entropy production increases as time and then saturates due to the finite-size effect. Using the probability distribution functions of σ, we calculate the average $\langle e^{-\sigma} \rangle$ of the integral fluctuation theorem as shown in Fig. 6.5b. These values are in good agreement with the values of $1 - \lambda = e^{-D_0}$.

6.4.3 Lower Bound for Irreversibility

In Fig. 6.6, the averaged entropy production $\langle\sigma\rangle$ is plotted against the time T in the logarithmic and normal scales. We see that the averaged entropy production $\langle\sigma\rangle$ is larger than the Rényi-0 divergence D_0. In a short time scale, the Rényi-0 divergence D_0 exponentially grows in time as

$$D_0 \simeq \frac{l}{w} e^{\Lambda T} \tag{6.21}$$

as shown in the green line in Fig. 6.6a. On the other hand, in an intermediate time scale, the Rényi-0 divergence D_0 grows linearly in time. This can be understood from the fractality $D_0 \simeq d_{\mathrm{C}} \ln[l/l_0]$ and the finest-structure length scale $l_0 = we^{-\Lambda T}$ as

$$D_0 \simeq d_{\mathrm{C}} \Lambda T + d_{\mathrm{C}} \ln \frac{l}{w}. \tag{6.22}$$

Thus, the transient fractality set the bound for the information-theoretic distance. The linear growth of this distance is reminiscent of the constant entropy production rate in steady states.

The empirical entropy production $\langle\sigma\rangle$ gives the information-theoretic distance between ρ_T and $\mathscr{C}_l[\rho_T]$, which is measured with nonzero accuracy l. As time proceeds, this distance becomes larger and the information about the true structure becomes lost to a significant degree. At the same time, the initial state $\tilde{\rho}_0^l$ that observers retrodict by their measurement becomes more different from the true initial state ρ_0. This information loss is controlled by the transient fractality in the intermediate time scale. This informational irreversibility is inevitable no matter how small

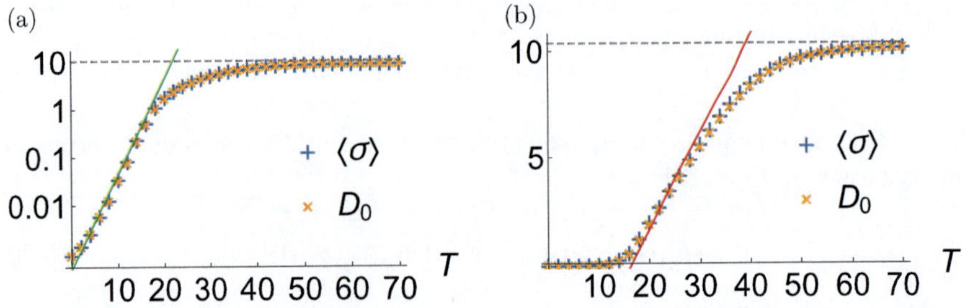

Fig. 6.6 Averaged entropy production $\langle\sigma\rangle$ and the Rényi-0 divergence D_0 **a** in the logarithmic scale and **b** in the normal scale. The blue marks represent the average $\langle\sigma\rangle$ calculated from the PDF of σ. The yellow marks represent the values of the Rényi-0 divergence. The grey dashed lines represent the limit due to a finite-size effect. The green line represents an exponential growth due to the Lyapunov exponent as $e^{\Lambda T}$, while the red line shows a linear growth due to the transient fractality as $d_{\mathrm{C}}\Lambda T$ + const. Reprinted figure from Ref. [1] under the Creative Commons Attribution 4.0 International license

a chosen length scale l is, because the transient fractality emerges in a large T due to chaoticity.

Appendix

A.1 Fractality and Rényi-0 Divergence

We here introduce the fractal dimension and discuss its relation to the Rényi-0 divergence.

A.1.1 Fractal Dimension

By way of simple illustration, let us consider a self-similar fractal named the Cantor set as depicted in Fig. 6.7. Initially, a line with a unit length denoted by C_0 is prepared. One third in the middle of the line is removed to obtain two disconnected segments C_1. Then, one third in the middle of each segment is removed to obtain four segments C_2. By repeating this procedure infinitely many times, we obtain the Cantor set C_∞.

The Cantor set has a self-similar property. In fact, the set obtained by the reduction of its size by the ratio $r = 1/3$ is congruent with its left or right part. Therefore, by the reduction with ratio $r = 1/3$, the Cantor set reduces its 'volume' by the factor of $f = 1/2$. Motivated by the fact that the volume reduction f and the size reduction r has the relation $f = r^d$ for ordinary objects like lines ($d = 1$), squares ($d = 2$) and cubes ($d = 3$), we define the dimension of the Cantor set by

$$d_{\mathrm{F}} = \frac{\ln f}{\ln r} = \frac{\ln 2}{\ln 3}, \tag{6.23}$$

which is called the fractal dimension for self-similar fractals.

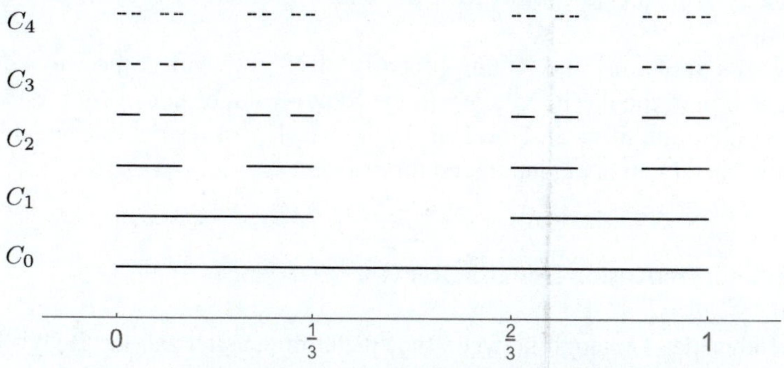

Fig. 6.7 Procedure to make the Cantor set. One third in the middle of each segment of C_n is removed to obtain C_{n+1}. Repeating this procedure infinite times, we obtain the Cantor set C_∞

We can extend this idea for more general set in the embedded space with dimension d_{E}. To this aim, we cover the set by a set of d_{E}-dimensional balls with diameter l. Let $N(l)$ denote the minimal number of balls needed to cover the set. Then, $l^{d_{\mathrm{F}}} N(l)$ ($\sim$ const.) should give a good estimate of the 'volume' of the set of interest as long as l is small enough. On the basis of this observation, we define the fractal dimension, or more precisely speaking, the box-counting dimension by [17]

$$d_{\mathrm{F}} = \lim_{l \to +0} \frac{\ln N(l)}{\ln(1/l)}. \tag{6.24}$$

Mathematically, we should take the limit of infinitesimal ε. However, in physical applications such as numerical simulations, we cannot generate infinitely small structures. Therefore, in the real world, the scaling (6.24) should break down around the smallest structure l_0. Hence, if the scaling relation is satisfied for sufficiently small l as

$$N(l) = N(l_0) \left(\frac{l_0}{l} \right)^{d_{\mathrm{F}}} \quad (l_0 \lesssim l \ll L), \tag{6.25}$$

we regard the structure as a physical fractal, where L is a typical extension of the fractal.

We can define the box-counting dimension without considering a probability distribution on the fractal. However, when the fractal has a probability on it, we can define other types of fractal dimensions. As such a generalization, the Rényi-α dimension is defined by

$$d_\alpha = \lim_{l \to +0} \frac{H_\alpha(l)}{\ln(1/l)}, \tag{6.26}$$

where H_α is the Rényi-α entropy defined by

$$H_\alpha(l) = \frac{1}{1-\alpha} \ln \sum_\gamma P(\gamma, l)^\alpha \tag{6.27}$$

with $P(\gamma, l)$ being the probability to be in a coarse-grained mesh with size l centered at a point γ.

In general situations, the scaling property of $P(\gamma, l)$ with respect to l depends on the position of the mesh. As a result, the Rényi-α dimensions vary for α and the fractal is called multifractal. However, in this thesis, we restrict our attention to a simple unifractal with a unique fractal dimension.

A.1.2 Fractal Dimension from the Rényi-0 Divergence

We here consider a relation between the fractal dimension and the Rényi-0 divergence. To be specific, let $\rho(\gamma)$ denote the probability distribution function in the embedding space. To see the vulnerability of this structure at a scale l, we convolute

ρ with the d_{E}-dimensional isotropic Gaussian with the standard deviation l as

$$\mathscr{C}_l[\rho](\gamma) := \int d\gamma\ \rho(\gamma') \left(\frac{1}{\sqrt{2\pi l^2}}\right)^{d_{\mathrm{E}}} \exp\left[-\frac{(\gamma-\gamma')^2}{2l^2}\right]. \tag{6.28}$$

In physical terms, this may be interpreted that a Gaussian noise is added to phase-space points in ρ to obtain $\mathscr{C}_l[\rho]$ at the ensemble level. We evaluate the difference between the original structure and the perturbed structure by the Rényi-0 divergence as

$$D_0(\rho||\mathscr{C}_l[\rho]) = -\ln \int_{\rho(\gamma)>0} d\gamma\ \mathscr{C}_l[\rho](\gamma). \tag{6.29}$$

The integral on the right-hand side represents how much fraction of $\mathscr{C}_l[\rho]$ remains in the original support after perturbation. Technically speaking, we can evaluate the integral by the Monte-Carlo method. To be specific, we choose a phase-space point in accordance with the original probability distribution ρ. Then, we add the Gaussian noise to it and check whether or not the obtained point is in the original support. If yes, this event contributes to the integral in Eq. (6.29). By repeating this procedure, we can numerically obtain the value of the Rényi-0 divergence D_0.

Suppose that we cover the fractal that satisfy (6.25) separately by d_{E}-dimensional spheres with the radius l and by those with the radius l_0. The number of spheres with the radius l_0 in a single sphere with the radius l should be $N(l_0)/N(l)$ as long as the spheres are uniformly distributed along the fractal.[2] Therefore, the probability that a phase-space point on the fractal stays on the fractal after perturbation with the length l is evaluated to be the ratio of the total volume of the spheres with the radius l_0 to the volume of the sphere with radius l as

$$\frac{N(l_0)}{N(l)} l_0^{d_{\mathrm{E}}} \Big/ l^{d_{\mathrm{E}}} = \left(\frac{l_0}{l}\right)^{d_{\mathrm{C}}} \quad (l_0 \lesssim l \ll L), \tag{6.30}$$

where we introduce the fractal codimension $d_{\mathrm{C}} := d_{\mathrm{E}} - d_{\mathrm{F}}$. Consequently, the integral in Eq. (6.29) is replaced with this quantity, leading to

$$D_0(\rho||\mathscr{C}_l[\rho]) = d_{\mathrm{C}} \ln \frac{l}{l_0}\ (l_0 \lesssim l \ll L). \tag{6.31}$$

Thus, the linear dependence of D_0 on $\ln l$ implies fractality and its proportionality factor coincides with the fractal codimension; the Rényi-0 divergence can be utilized to evaluate the fractal dimension.

Although Eq. (6.31) can heuristically be derived as seen above, it is yet to be proven rigorously. Nevertheless, its validity can numerically be confirmed in simple fractals as shown below.

---

[2] Here, we use the assumption of unifractality.

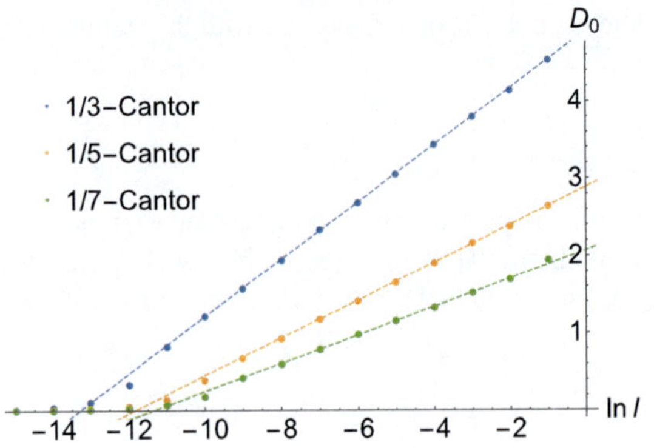

Fig. 6.8 Rényi-0 divergence of the Cantor sets. The Rényi-0 divergence is numerically calculated against the variation of l by Monte-Carlo simulations for three tenth-level Cantor sets C_{10}. The dashed lines are the guides to the eye with their slopes giving the corresponding fractal codimensions. To obtain each data point, 10^4 states are generated

Cantor sets

We numerically evaluate the Rényi-0 divergence for three types of the Cantor sets as shown in Fig. 6.8. The $1/k$-Cantor set is the Cantor set obtained by the repeated removals of one kth in the middle of each segment. The fractal dimension of the $1/k$-Cantor set is evaluated to be

$$d_{\mathrm{F}} = \frac{\ln 2}{\ln \frac{2k}{k-1}}. \tag{6.32}$$

The Rényi-0 divergence D_0 is numerically evaluated by the Monte-Carlo method. We can see that D_0 is proportional to $\ln l$ as expected. The slopes of the lines are consistent with the corresponding values of the fractal codimension.

Sierpinski gasket

We can apply the method to higher-dimensional cases. We here consider the Sierpinski gasket shown in Fig. 6.9a. From the argument based on the self-similarity, we can obtain the fractal dimension

$$d_{\mathrm{F}} = \frac{\ln 3}{\ln 2}. \tag{6.33}$$

The Rényi-0 divergence D_0 is numerically calculated as shown in Fig. 6.9b. The proportionality relation between D_0 and $\ln l$ is confirmed for intermediate values of l. The proportionality constant is consistent with the value of the fractal codimension $d_{\mathrm{C}} = 2 - \ln 3/\ln 2$.

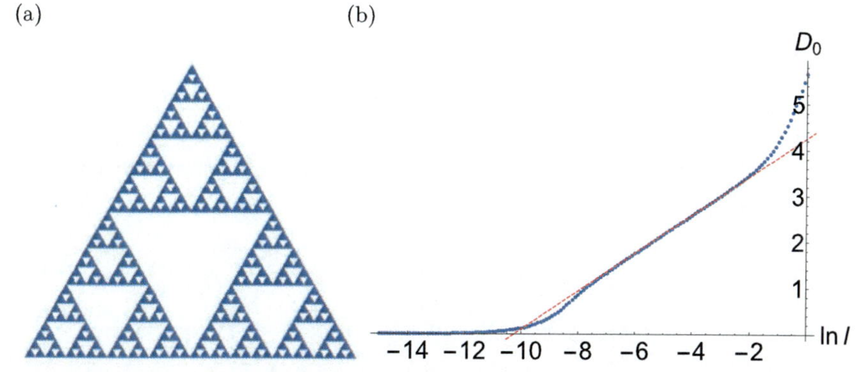

Fig. 6.9 Sierpinski gasket and its Rényi-0 divergence. **a** Sierpinski gasket. The Sierpinski gasket can be obtained by the repeated removals of the equilateral triangle at the center. **b** Rényi-0 divergence obtained from numerical simulations. The blue dots shows numerically obtained values of D_0. The red dashed line is a guide to the eye with the slope $d_C = 2 - \ln 3/\ln 2$. They agree with each other excellently for intermediate values of l. To obtain each data point, 10^5 Monte-Carlo samples are taken

A.2 Mathematical Support of the Transient Fractality

We here introduce a method by Kaplan and Yorke to evaluate the fractal dimension of the system from the Lyapunov spectrum [17]. Applying this method, we conjecture that a generic chaotic Hamiltonian system with the spatial dimension d has the transient fractal with dimension $d_F = d$.

A.2.1 Kaplan-Yorke Conjecture

Let us denote the Lyapunov exponents in the nonincreasing order as

$$\Lambda_1 \geq \Lambda_2 \geq \cdots \geq \Lambda_{d_E}. \tag{6.34}$$

Then, we define

$$\xi_i = \sum_{j=1}^{i} \Lambda_j. \tag{6.35}$$

Let k be the largest value that satisfies $\xi_k \geq 0$. Then the Lyapunov dimension of the system is defined by [17]

$$d_L = k - \frac{\xi_k}{\Lambda_{k+1}}. \tag{6.36}$$

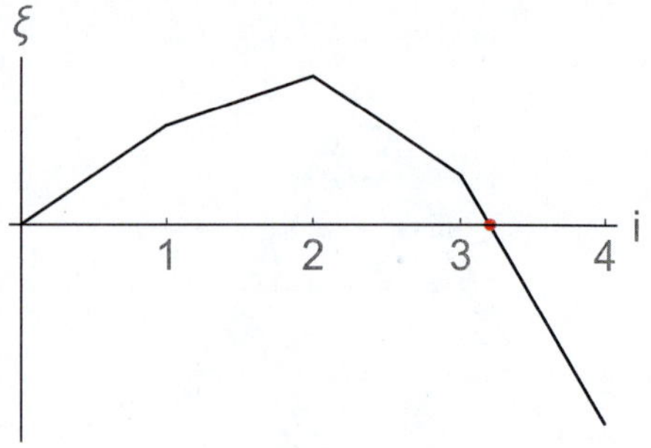

Fig. 6.10 Evaluation of the Lyapunov dimension. The values of ξ_i are plotted against i and adjacent points are connected by lines. The point at which the line goes below the i axis for the first time represents the value of the Lyapunov dimension depicted as the red dot

It is conjectured that this Lyapunov dimension gives a good evaluation of the fractal dimension[3] [17]. We can schematically represent the value of d_L as shown in Fig. 6.10. We plot the values of ξ_i with lines interpolating between adjacent points. The crossing point of these lines with the i axis indicates the value of the Lyapunov dimension d_L.

A.2.2 Conjecture for Higher-Dimensional Cases

Let us consider the Hamiltonian system with the spatial dimension of d. Due to the conservation of energy, we can represent phase space as a d_E-dimensional manifold with $d_E = 2d - 1$. From the symplectic property of the Hamiltonian system, the Lyapunov exponents has the symmetry

$$\Lambda_1 = -\Lambda_{2d-1},\ \Lambda_2 = -\Lambda_{2d-2}, \ldots, \Lambda_{d-1} = -\Lambda_{d+1},\ \Lambda_d = -\Lambda_d = 0. \quad (6.37)$$

By applying the Kaplan-Yorke method as shown in Fig. 6.11a, we obtain the dimension $2d - 1$. Therefore, the dimension of a phase-space object does not change by the dynamics. This result is consistent with the fact that we have no fractal in the long-time limit.

Meanwhile, in the transient time scale, positive Lyapunov exponents are expected not to play significant roles since they do not affect local structures in phase space. Therefore, the Lyapunov spectrum effectively reduces to

$$\{0, 0, \ldots, 0, \Lambda_{d+1}, \Lambda_{d+2}, \ldots, \Lambda_{2d-1}\}. \quad (6.38)$$

By applying the Kaplan-Yorke method to this Lyapunov spectrum, we obtain the fractal dimension of $d_F = d$. Hence, we conjecture that a fractal structure with dimension $d_F = d$ transiently emerges in a chaotic Hamiltonian system with the spatial dimen-

[3] The fractal dimension discussed here is actually the dimension of the natural measure and a concept generally different from the fractal dimension that we have introduced in the previous section [17]. However, we naively assume that the Lyapunov dimension typically gives an estimate of our fractal dimension as well.

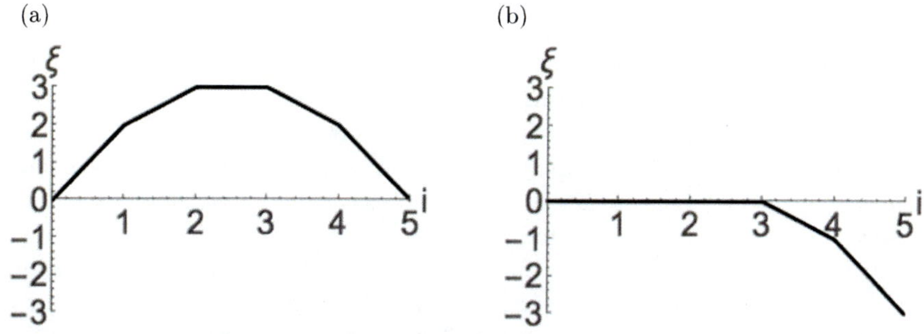

Fig. 6.11 Evaluations of the fractal dimension by the Kaplan-Yorke method. **a** Evaluation for the long-time limit. The resulting value $2d - 1$ is the same as the dimension of the embedding space. Therefore, no fractality emerges in the long-time limit. **b** Evaluation for the transient time regime. The obtained value $d_F = d$ implies that transient fractal emerges with this dimension. We here consider the case of $d = 3$ for simplicity of illustration

sion d. Note that the numerical simulations in the main text demonstrate the validity of this conjecture for $d = 2$.

A.3 Details of Numerics

A.3.1 Evaluation of the Rényi Divergence

As we have discussed in Sect. 6.3, we evaluate the Rényi-0 divergence by utilizing the time reversal test. We use the Monte-Carlo method to evaluate the integral in the Rényi-0 divergence. To be specific, we sample a phase-space point γ_0 from the initial uniform distribution ρ_0 over a cube. In this sampling, we use a low-discrepancy sequence generated by the additive recurrence to improve the precision of numerical simulations. Then, we let the state evolve over time T and obtain the corresponding final point γ_T, whose ensemble average is ρ_T. Next, we add a white Gaussian noise with the standard deviation l to the time-reversal of this phase-space point $\mathscr{T}\gamma_T$ to obtain $\tilde{\gamma}_T$. When the obtained point happens to be outside the stadium, we add a different noise to $\mathscr{T}\gamma_T$ until $\tilde{\gamma}_T$ falls inside the stadium. In other words, we truncate the Gaussian so as not to obtain unphysical points. The ensemble of $\tilde{\gamma}_T$ constitutes $\tilde{\rho}_T$ except for an insignificant deviation due to this truncation. Finally, we let $\tilde{\gamma}_T$ evolve over time T to obtain $\tilde{\gamma}_0$. We judge whether or not its time reversal $\mathscr{T}\tilde{\gamma}_0$ is in the support of the original state ρ_0. Iff yes, this sample contributes to the integral in the Rényi-0 divergence.

In Fig. 6.4, the length scale l is increased from e^{-20} to 1 by the multiplication of $e^{0.2}$; the time T is varied from 0 to 70 by the increment of 2. For each fixed l and T, we collect 10^9 samples to evaluate the Rényi-0 divergence D_0. The numerical precision becomes worse as the value of D_0 increases. Nevertheless, the relative statistical error is less than two percent in our simulations. We calculate the value of the derivative $dD_0/d(\ln l)$ by locally applying the least square fitting to D_0 with respect to $\ln l$. The fitting is done over eleven adjacent data points that consist of the point of our interest, five points on its left and five on its right.

To obtain the probability distribution of σ as in Fig. 6.5, we need more detailed information than we do in the case of D_0. To this aim, we divide the cube of the initial state into 8^3 sub-cubes. Then, from each sub-cube we collect $2^{24} \sim 2 \cdot 10^7$ samples as the initial state and conduct the time reversal test. Therefore, the total sampling number is $2^{33} \sim 9 \cdot 10^9$. The number of the states that terminate in each sub-cube is calculated to obtain the probability distribution function $\tilde{\rho}_0$ in the original cube. The time T is varied from 0 to 70 by the increment of 2. As time proceeds, the number of states that terminate in each sub-cube decreases since $\tilde{\rho}_0$ diffuses. Even in this case, the number is typically larger than a few hundred and therefore the relative statistical errors are at most five percent.

A.3.2 Chaos and Numerical Precision of Floating Numbers

Since we deal with a chaotic system, we should be careful about the numerical precision of floating numbers. To see the effect of the precision, we conduct the time reversal test with no noise added. In this case, the difference between the initial state and the pullbacked state is caused by the numerical limitation. Let p denote the numerical precision, namely, 10^{-p} represents the absolute precision of floating numbers. Then, the largest error in the time reversal test is caused by the error in the stable direction at the final state. When the time reversal test is performed, this error grows by the exponential factor $e^{\Lambda T}$, since the stable direction changes into the unstable direction upon the time reversal. Therefore, the typical error for the time reversal test without noise is expected to be

$$\varepsilon \simeq 10^{-p} e^{\Lambda T} \simeq 10^{0.2T-p} \tag{6.39}$$

for our Bunimovich billiard with $\Lambda = 0.46$.

We conduct the time reversal test without noise while varying the numerical precision p. By repeating the time reversal test with different initial conditions under a fixed T, we can obtain the distribution of the distance between the initial state and the pullbacked state in phase space. The median of this distance $d_{\rm m}$ is plotted in Fig. 6.12a. The reason why we do not use the average is that the distribution is heuristically similar to the log-normal distribution and therefore the average does not represent a typical value. We can see that the median is almost equal to ε. We also calculate the change in the median $d_{\rm m}$ against the variation of the evolution time T as shown in Fig. 6.12b. Again, the result is consistent with Eq. (6.39).

Hence, for our numerical simulations to be reliable, the parameters should satisfy $w \gg \varepsilon \simeq 10^{0.2T-16}$, since we use double precision numbers in the numerical simulations in the main text. Recall that w is the side of the initial cube. Therefore, for Fig. 6.4 with $w = 0.01$, the reliability is guaranteed for $T \lesssim 70$. Thus, the data

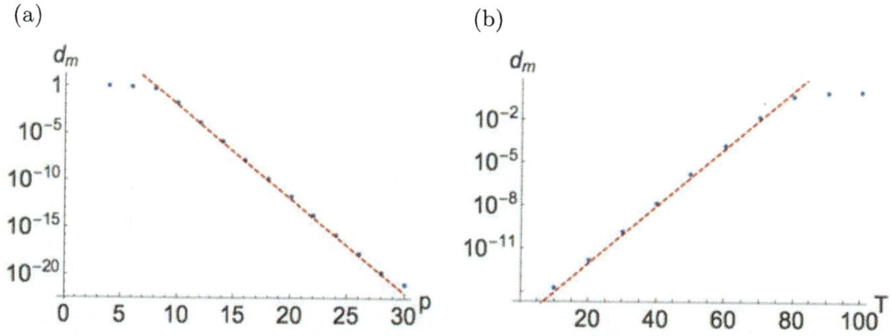

Fig. 6.12 Errors caused by the numerical precision. **a** Dependence of the distance d_{m} on the numerical precision p. The median of the distance (blue dots) obtained by the time reversal test without noise is plotted against different values of the numerical precision p. The red dashed line represents the theoretical estimate of typical errors (6.39). The evolution time is fixed to be $T = 40$. **b** Dependence of the distance d_{m} on the evolution time T. The median of the distance (blue dots) are depicted for different time T. The red dashed line represents Eq. (6.39). The precision is fixed to $p = 16$, namely, the double precision. Reprinted figure from Ref. [1] under the Creative Commons Attribution 4.0 International license

plotted in Fig. 6.4 are reliable despite numerical difficulty of simulating the chaotic system.

References

1. Murashita Y, Kura N, Ueda M (2021) Phys Rev Res 3:033224
2. Boltzmann L (1872) Sitzungsber Kais Akad Wiss Wien, Math-Naturwiss Cl Abt II 66:275
3. Loschmidt J (1876) Sitzungsber Kais Akad Wiss Wien, Math-Naturwiss Cl Abt II 73:128
4. Boltzmann L (1877) Sitzungsber Kais Akad Wiss Wien, Math-Naturwiss Cl Abt II 75:62
5. Holian BL, Hoover WG, Posch HA (1987) Phys Rev Lett 59:10
6. Evans DJ, Cohen EGD, Morriss GP (1993) Phys Rev Lett 71:2401
7. Gallavotti G, Cohen EGD (1995) Phys Rev Lett 74:2694
8. Gallavotti G, Cohen EGD (1995) J Stat Phys 80:931
9. Gallavotti G (2014) Nonequilibrium and irreversibility. Springer International Publishing
10. Nosé S (1984) Mol Phys 52:255
11. Hoover WG (1985) Phys Rev A 31:1695
12. Hoover WG, Hoover CG (2012) Time reversibility, computer simulation, algorithms, chaos, 2nd edn. Advanced series in nonlinear dynamics. World Scientific Publishing Company
13. van Erven T, Harremoës P (2014) IEEE Trans Inf Theory 60:3797
14. Bunimovich LA (1979) Commun Math Phys 65:295
15. Shepelyansky DL (1983) Physica 8D:208
16. Adachi S, Toda M, Ikeda K (1988) Phys Rev Lett 61:659
17. Farmer JD, Ott E, Yorke JA (1983) Physica 7D:153
18. Umberger DK, Farmer JD (1985) Phys Rev Lett 55:661
19. Grebogi C, McDonald SW, Ott E, Yorke JA (1985) Phys Lett 110A:1

Chapter 7
Summary and Outlook

7.1 Summary

We have studied the fluctuation theorems with divergent entropy production, namely, the ones with absolute irreversibility and the ones in the overdamped theory in the presence of multiple heat reservoirs. We have also discussed the application of the former to two fundamental problems in statistical physics, namely, the Gibbs paradox and the Loschmidt paradox.

In Chap. 2, we have briefly reviewed history of the fluctuation theorems. The theorems indicate that the probability of a negative entropy production is exponentially suppressed compared with that of the positive counterpart. Their outstanding feature is that they apply to a wide range of nonequilibrium systems beyond a linear-response regime. They were initially conjectured by the invariant measure of a dissipative chaotic system and demonstrated in numerical simulations of a shear-driven flow. Subsequently, they were shown under various dynamics and both in steady-state and transient situations under time-independent driving. Then, nonequilibrium work relations including the Jarzynski equality and the Crooks fluctuation theorem were discovered under time-dependent driving. Afterward, similar fluctuation theorems for various kinds of entropy productions were found. These fluctuation theorems have a common structure in that the ratio of the probability distribution function in the physical process to that in the reference process can be expressed by the corresponding entropy production.

In Chap. 3, we have reviewed the fluctuation theorems in the presence of absolute irreversibility. Despite the wide applicability of the fluctuation theorem, they cannot be applied to free expansion. We have seen that this inapplicability is rooted in absolute irreversibility, which physically corresponds to negatively divergent entropy production and mathematically originates from the singularity of the probability measure in the reference process with respect to that in the original process. By separating this singularity from the ordinary irreversible part by the Lebesgue decomposition theorem, we have derived the fluctuation theorems with absolute irreversibility. The degree of absolute irreversibility can ordinarily be calculated as the sum of two contributions. One is the singular continuous contribution given as the sum of the probability of the reference paths that do not have counterparts in the original pro-

Y. Murashita, *Fluctuation Theorems under Divergent Entropy Production and their Applications for Fundamental Problems in Statistical Physics*, Springer Theses, https://doi.org/10.1007/978-981-16-8638-2_7

cess. The other is the discrete contribution given as the sum of the probability of the reference paths with δ-function-type localizations. We have demonstrated the validity of the theorems in some examples.

In Chap. 4, we have studied the fluctuation theorems in the presence of multiple heat reservoirs. In this system, naive overdamped approximations fail since the velocity degrees of freedom relax to a nonequilibrium steady state and therefore have positively divergent contributions to the entropy production. To construct a proper overdamped approximation, we have gone back to the underdamped description and shown the underdamped fluctuation theorems. Then, by applying the singular expansion to the time-evolution equation of the generating function of thermodynamic quantities, we have derived the correct overdamped approximation, which separates the dynamics of the positional degrees of freedom from underdamped dynamics in the limit of infinitesimal velocity relaxation time. We have shown that the fluctuation theorems are satisfied for the overdamped contributions from the positional degrees of freedom, although the velocity degrees of freedom give singular contributions to the entropy production.

In Chap. 5, we have considered the relation between the Gibbs paradox and the fluctuation theorem with absolute irreversibility. The Gibbs paradox started from his contemplation of foundations of thermodynamics and statistical mechanics by means of gas mixing and now refers to a collective set of related but distinct problems. Following van Kampen, we have classified the Gibbs paradox into three aspects and seen that all the aspects had completely been understood within classical theory in the thermodynamic limit. Among them, we have revisited the issue of how to determine the relation between the thermodynamic entropy and the statistical-mechanical entropy. In the thermodynamic limit, this issue is resolved by the requirement of extensivity for the thermodynamic entropy as Jaynes pointed out. However, this resolution breaks down in a small thermodynamic system since extensivity itself breaks down. We have shown that the fluctuation theorem with absolute irreversibility takes the place of extensivity, that is, by requiring the theorem we can determine the relation between the entropies. Remarkably, this procedure is applicable to an interacting system as long as additivity holds.

In Chap. 6, we have revisited the Loschmidt paradox in the context of the fluctuation theorems. The Loschmidt paradox states that irreversible macroscopic behaviors cannot arise from reversible equations of motion due to the one-to-one correspondence between a path with a positive entropy production and that with its sign-reversed entropy production. Boltzmann argued that despite this correspondence the occurrence of paths with positive entropy productions can be overwhelmingly more frequent. In dissipative systems, fractality plays key roles in demonstrating this argument numerically or quantitatively by means of the steady-state fluctuation theorem. However, this fractal explanation of emergent irreversibility cannot naively be applied to Hamiltonian systems since the phase-volume conservation prohibits fractality in the long-time limit. Nevertheless, we have shown that a chaotic Hamiltonian system exhibits fractality in a transient time scale when an initial state is small compared to the system size. To evaluate the fractality, we have invoked a time reversal test by an imperfect Loschmidt demon. By regarding the imperfect backward process as the reference process, we have formulated the fluctuation theorem with absolute irreversibility. From the theorem, we have shown that the informational irreversibility is bounded in terms of the fractal dimension of the phase-space structure in the transient time scale. The linear relation between the bound and the evolution time is reminiscent of the stationary entropy production with a constant rate in a steady state.

7.2 Our Related Work

The fluctuation theorems can be generalized to situations under feedback control like Maxwell's demon [1, 2]. Under feedback control, the state of the system is measured and the parameters of the system are controlled by means of the information obtained by the measurement. In such situations, the information content can be treated on equal footing with the entropy production and hence the fluctuation theorems can be derived [3, 4]. When we conduct an error-free measurement, the possible paths of the forward process are restricted to a small subset of path space. Moreover, when the feedback operation is non-optimal, backward paths generically diffuse to the entire path space. Therefore, we have backward paths that lack the counterparts in the forward process. Thus, under feedback control, absolute irreversibility naturally emerges. We have derived the fluctuation theorem even under such situations [5]. Moreover, absolute irreversibility can be interpreted as inevitable information loss due to non-optimality of the feedback operation [6]. The derived fluctuation theorem can be used to optimize a given set of parameters used in the feedback operation [6].

The fluctuation theorems can be derived in quantum systems [7, 8], where the total system consisting of the system of our interest and the heat reservoir undergoes a unitary evolution. These quantum fluctuation theorems can be extended to situations under feedback control [9]. Under feedback control based on a projective measurement, absolute irreversibility again emerges due to the same reason as the error-free measurement in classical cases. Hence, we have modified the fluctuation theorems to include the degree of absolute irreversibility [10].

In open quantum systems governed by the Lindblad equation, the fluctuation theorems have been derived on the basis of the quantum trajectory approach [11–13], extended under feedback control [14], and generalized in the presence of absolute irreversibility [15]. Remarkably, absolute irreversibility captures thermodynamic advantage of quantum coherent driving. The coherent driving helps the state of the system diffuse in state space and therefore mitigates the restriction on the forward process. Consequently, absolute irreversibility is suppressed by the coherent driving, which implies that the second-law-like inequality derived from the fluctuation theorem is weaker under the coherent driving. Meanwhile, in the absence of the coherent driving, absolute irreversibility emerges and gives inevitable dissipation.

The classical fluctuation theorems under feedback control have experimentally verified in a colloidal system [16] and a single-electron box [17]. By using a superconducting qubit, we have verified the quantum fluctuation theorems under feedback control [18]. Under feedback control based on a projective measurement, we have seen that absolute irreversibility actually emerges, giving restriction of extractable work. Meanwhile, with a weak measurement, absolute irreversibility disappears and the average of the integral fluctuation theorem takes the value of unity except for influences of unwanted relaxation of a qubit.

7.3 Outlook

Recently, efficiency of small engines has attracted renewed interest [19–22]. The overdamped approximation in the presence of multiple heat reservoirs derived in Chap. 4 is expected to be useful to analytically study efficiency of heat engines in a steady state.

In Refs. [23, 24], a Langevin system coupled to two heat reservoirs with different temperatures is experimentally realized in an electrical circuit to verify the underdamped fluctuation theorems. Our overdamped theory and the overdamped fluctuation theorems can experimentally be tested in a similar setup with suitable parameters that guarantee the fast momentum relaxation.

In Chap. 5, we have considered the Gibbs paradox under the assumption of additivity of the entropies. However, it is known that additivity is no longer valid in systems with long-range interactions [25]. It might be interesting to see whether we can extend our approach to systems with nonadditive entropy.

We have seen that quantum statistical mechanics has the ambiguity function as classical statistical mechanics does. Hence, this ambiguity should be removed by a thermodynamic requirement for the entropy, as we have done by requiring the fluctuation theorem with absolute irreversibility in the classical case. A naive extension seems difficult since the work upon gas mixing fluctuates because of the discrete energy levels in contrast to the classical case. Therefore, seeking for the requirement suitable for the quantum case may be an interesting issue.

The difference between our ambiguity function and that of the Pauli-Jaynes resolution can be significant when we evaluate chemical potentials in stochastic thermodynamics of open chemical reaction networks with small particle numbers (see, e.g., [26]), and may therefore be experimentally detectable.

In Chap. 6, we have bounded the informational irreversibility by fractality. It may be natural that this informational irreversibility is related to the thermodynamic irreversibility. In particular, it may worth investigating the relation between the empirical entropy production formally defined by the fluctuation theorem and the thermodynamic entropy.

Classically, the dynamics can create any tiny phase-space structure. However, when we consider quantum systems, the length scale corresponding to the Planck constant manifests itself. Therefore, the smallest structure cannot be determined only by the Lyapunov exponents. Hence, interplay between the Planck constant and the Lyapunov exponents may cause interesting effects on irreversibility in quantum systems.

In this thesis, we have seen that the fluctuation theorems apply to genuinely nonequilibrium situations with divergent entropy production when we take into account absolute irreversibility. Furthermore, we have shown that the fluctuation theorems have considerable implications for the foundation of statistical physics. The last question that I would like to pose is whether we can reconstruct operational thermodynamics by requiring the fluctuation theorem with absolute irreversibility as an axiom. Actually, in Chap. 5, we have demonstrated that we can define the free-energy difference and the entropy production by axiomatically requiring the Jarzynski equality with absolute irreversibility. If this requirement were enough to characterize thermodynamics, equilibrium thermodynamics, steady-state thermodynamics and thermodynamics for slow degrees of freedom might be formulated in a unified way by axiomatizing the absolutely irreversible fluctuation theorems, i.e., the Jarzynski equality, the Hatano-Sasa equality and the fluctuation theorems in Chap. 4, respectively. Furthermore, from the empirical entropy production and the operational definition of absolute irreversibility in the time reversal test in Chap. 6, we could construct thermodynamics for isolated systems. If these speculations were true, the fluctuation theorems would lay the foundation for statistical physics in a genuine sense.

References

1. Maxwell JC (1871) Theory of heat. Appleton
2. Sagawa T (2012) Thermodynamics of information processing in small systems. Springer
3. Sagawa T, Ueda M (2010) Phys Rev Lett 104:090602
4. Horowitz JM, Vaikuntanathan S (2010) Phys Rev E 82:061120
5. Murashita Y, Funo K, Ueda M (2014) Phys Rev E 90:042110
6. Ashida Y, Funo K, Murashita Y, Ueda M (2014) Phys Rev E 90:052125
7. Tasaki H (2000) arXiv:cond-mat/0009244
8. Campisi M, Hänggi P, Talkner P (2011) Rev Mod Phys 83:1653
9. Funo K, Watanabe Y, Ueda M (2013) Phys Rev E 88:052121
10. Funo K, Murashita Y, Ueda M (2015) New J Phys 17:075005
11. Horowitz JM (2012) Phys Rev E 85:031110
12. Horowitz JM, Parrondo JMR (2013) New J Phys 15:085028
13. Hekking FWJ, Pekola JP (2013) Phys Rev Lett 111:093602
14. Gong Z, Ashida Y, Ueda M (2016) Phys Rev A 94:012107
15. Murashita Y, Gong Z, Ashida Y, Ueda M (2017) Phys Rev A 96:043840
16. Toyabe S, Sagawa T, Ueda M, Muneyuki E, Sano M (2010) Nat Phys 6:988
17. Koski JV, Maisi VF, Sagawa T, Pekola JP (2014) Phys Rev Lett 113:030601
18. Masuyama Y, Funo K, Murashita Y, Noguchi A, Kono S, Tabuchi Y, Yamazaki R, Ueda M, Nakamura Y (2018) Nat Commun 9:1291
19. Verley G, Esposito M, Willaert T, Van den Broeck C (2014) Nat Commun 5:4721
20. Verley G, Willaert T, Van den Broeck C, Esposito M (2014) Phys Rev E 90:052145
21. Polettini M, Verley G, Esposito M (2015) Phys Rev Lett 114:050601
22. Park JM, Chun HM, Noh JD (2016) Phys Rev E 94:012127
23. Ciliberto S, Imparato A, Naert A, Tanase M (2013) Phys Rev Lett 110:180601
24. Ciliberto S, Imparato A, Naert A, Tanase M (2014) J Stat Mech: Theor Exp P1:2013
25. Tsallis C (2009) Introduction to nonextensive statistical mechanics. Springer
26. Rao R, Esposito M (2016) Phys Rev X 6:041064